Six Lectures on Light

Delivered in America in 1872–1873

JOHN TYNDALL

CAMBRIDGE
UNIVERSITY PRESS

CAMBRIDGE UNIVERSITY PRESS

Cambridge, New York, Melbourne, Madrid, Cape Town,
Singapore, São Paolo, Delhi, Tokyo, Mexico City

Published in the United States of America by Cambridge University Press, New York

www.cambridge.org
Information on this title: www.cambridge.org/9781108038430

© in this compilation Cambridge University Press 2011

This edition first published 1873
This digitally printed version 2011

ISBN 978-1-108-03843-0 Paperback

CAMBRIDGE LIBRARY COLLECTION

Books of enduring scholarly value

Physical Sciences

From ancient times, humans have tried to understand the workings of
the world around them. The roots of modern physical science go back to
the very earliest mechanical devices such as levers and rollers, the mixing
of paints and dyes, and the importance of the heavenly bodies in early
religious observance and navigation. The physical sciences as we know them
today began to emerge as independent academic subjects during the early
modern period, in the work of Newton and other 'natural philosophers',
and numerous sub-disciplines developed during the centuries that followed.
This part of the Cambridge Library Collection is devoted to landmark
publications in this area which will be of interest to historians of science
concerned with individual scientists, particular discoveries, and advances in
scientific method, or with the establishment and development of scientific
institutions around the world.

Six Lectures on Light

Born in Leighlinbridge in Ireland, John Tyndall (1820–93) was a brilliant
nineteenth-century experimental physicist and gifted science educator. He
worked initially as a draughtsman, then spent a year teaching at an English
school before attending the University of Marburg to study physics and
chemistry. Tyndall carried out important research on magnetism, light and
bacteriology. Among his many significant achievements, he demonstrated
the greenhouse effect in Earth's atmospheric gases using absorption
spectroscopy. He was a skilled and entertaining educator and as Professor
of Natural Philosophy at the Royal Institution he gave many public lectures
and demonstrations of science. Published in 1873, this book features six
accessible lectures on light. They explore a wide range of ideas in a non-
technical way, from basic scientific theories through magnetism and light
scattering, to analytical spectroscopy. The book ends with a series of essays
on special topics, and includes a detailed index.

Cambridge University Press has long been a pioneer in the reissuing of out-of-print titles from its own backlist, producing digital reprints of books that are still sought after by scholars and students but could not be reprinted economically using traditional technology. The Cambridge Library Collection extends this activity to a wider range of books which are still of importance to researchers and professionals, either for the source material they contain, or as landmarks in the history of their academic discipline.

Drawing from the world-renowned collections in the Cambridge University Library, and guided by the advice of experts in each subject area, Cambridge University Press is using state-of-the-art scanning machines in its own Printing House to capture the content of each book selected for inclusion. The files are processed to give a consistently clear, crisp image, and the books finished to the high quality standard for which the Press is recognised around the world. The latest print-on-demand technology ensures that the books will remain available indefinitely, and that orders for single or multiple copies can quickly be supplied.

The Cambridge Library Collection will bring back to life books of enduring scholarly value (including out-of-copyright works originally issued by other publishers) across a wide range of disciplines in the humanities and social sciences and in science and technology.

LIGHT.

HANHART. LITH.

PLUMES PRODUCED BY THE CRYSTALLIZATION OF WATER.

Photographed by Professor Lockett.

SIX LECTURES ON LIGHT

DELIVERED IN AMERICA IN 1872–1873

BY

JOHN TYNDALL, LL.D. F.R.S.

PROFESSOR OF NATURAL PHILOSOPHY IN THE ROYAL INSTITUTION.

LONDON:

LONGMANS, GREEN, AND CO.

1873.

PREFACE

TO

THE ENGLISH EDITION.

———•◇•———

SINCE my return from the United States I have sought, by additions and emendations of various kinds, to render these Lectures more useful to my readers on both sides of the Atlantic.

JOHN TYNDALL.

ROYAL INSTITUTION :
June 1873.

PREFACE.

My eminent friend Professor JOSEPH HENRY, of the Smithsonian Institution, Washington, did me the honour of taking these lectures under his personal direction, and of arranging the times and places at which they were to be delivered.

Believing that my home duties could hardly be suspended for a longer period, I did not, at the outset, expect to be able to prolong my visit to the United States beyond the end of 1872.

Thus limited as to time, Professor HENRY began in the North, and, proceeding southwards, arranged for the successive delivery of the lectures in Boston, New York, Philadelphia, Baltimore, and Washington.

By this arrangement, which circumstances at the time rendered unavoidable, the lectures in New York would have been rendered coincident with the period of the presidential election. This was deemed unsatisfactory, and the fact being represented to me, I at once offered to extend the time of my visit so as to make the lectures in New York succeed those in Washington. The proposition was cordially accepted by my friends.

To me personally this modified arrangement has proved both pleasant and beneficial. It gave me a

much-needed and delightful holiday at Niagara Falls ;
it, moreover, rendered the successive stages of my
work a kind of *growth*, which reached its most impressive development in New York and Brooklyn.

My reception throughout has been that of a friend
by friends ; and now that my visit has become virtually
a thing of the past, I look back upon it as a memory
without a single stain of unpleasantness. Excepting
one inexorable event, nothing has occurred that I could
wish not to have occurred ; while from beginning to
end I have been met by expressions of good-will on the
part of the American people never, on my part, to be
forgotten.   Indeed, ' good-will ' is not the word to express the kindness manifested towards me in the United
States.

Would that it had been in my power to meet the wishes
of my friends more completely, by responding to the invitations sent to me from the great cities of the Interior and
the West.   But the character of the lectures, and their
weight of instrumental appliances, involved such heavy
labour that the need of rest alone would be a sufficient
reason for my pausing here.   Besides this, each successive mail from London brings me intelligence of work
suspended and duties postponed through my absence.

The Royal Institution possesses an honorary secretary
who has devoted the best years of an active professional
life and the best energies of a strong man to its
interests.   And if anything of the kind should ever be
founded here, the heartiest wish that I could offer for
its success would be, that it may be served with the
singleness of purpose, and self-sacrificing love, bestowed

by its managers and its members on the Royal Institution; and by none more unceasingly than by Dr. BENCE
JONES. But he, on whom I might rely, is now smitten
down by a distressing illness;[1] and, though others are
willing to aid me in all possible ways, there can be no
doubt as to my line of duty. I ought to be at home.
I ask my friends in the Interior and the West to take
these things into consideration; and to think of me not
as one insensible to their kindness, but as one who, with
a warmth commensurate with their own, would comply
with all their wishes if he could.

One other related point deserves mention. On quitting England I had no intention of publishing these
lectures, and, except a fragment or two, they were wholly
unwritten when I arrived in this city. Since that time,
besides lecturing in New York, Brooklyn, and New
Haven, the lectures have been written out and carried
through the Press. No doubt many evidences of the
rapidity of their production will appear; but I thought
it due to those who listened to them with such un-
wavering attention, as also to those who wished to hear
them, but were unable to do so, to leave them behind me
in an approximately authentic form. The constant
application which this work rendered necessary has
cut me off from many social pleasures; it has prevented me from making myself acquainted with the
working of institutions in which I feel a deep interest,
and from availing myself of the generous hospitality
offered to me by the clubs of New York. In short, it

---

[1] He died, working for the Institution to the last, on Sunday morning,
April 20, 1873.

has made me an unsociable man. But, finding social
pleasure and hard work incompatible, I took the line
of devoting such energy as I could command, not to
the society of my intimate friends alone, but to the
people of the United States.

In the opening lecture are mentioned the names of
gentlemen to whom I am under lasting obligations for
their friendly and often laborious aid. The list might
readily be extended, for in every city I have visited
willing helpers were at hand. I must not, however,
omit the name of Mr. RHEES, Professor HENRY's private
secretary, who not only in Washington, but in Boston,
gave me most important assistance. To the Trustees
of the Cooper Institute my acknowledgments are due;
and to the Directors of the Mercantile Library at
Brooklyn. I would add to these a brief but grateful
reference to my high-minded friend and kinsman
General HECTOR TYNDALE, for his long-continued care of
me, and for the thoughtful tenderness by which he and
his family softened, both to me and to the parents of
the youth, the pain occasioned by the death of my
junior assistant in Philadelphia.

Finally, I have to mention with warm commendation
the integrity, ability, and devotion with which, from
first to last, I have been aided by my principal assistant,
Mr. JOHN COTTRELL.

JOHN TYNDALL.

NEW YORK : *February* 1873.

CONTENTS.

LECTURE VI.

APPENDIX.

ON LIGHT.

LECTURE I.

SOME twelve years ago I published, in England, a
little book entitled the 'Glaciers of the Alps,' and,
a couple of years subsequently, a second volume, entitled 'Heat as a Mode of Motion.' These volumes were
followed by others, written with equal plainness, and with
a similar aim, that aim being to develop and deepen
sympathy between science and the world outside of
science. I agreed with thoughtful men[1] who deemed
it good for neither world to be isolated from the other,
or unsympathetic towards the other, and, to lessen
this isolation, at least in one department of science, I

[1] Among whom may be mentioned, especially, the late Sir Edmund
Head, Bart., with whom I had many conversations on this subject.

B

swerved aside from those original researches which had
previously been the pursuit and pleasure of my life.

The works here referred to were, for the most part,
republished by the Messrs. Appleton of New York,[1]
under the auspices of a man who is untiring in his
efforts to diffuse sound scientific knowledge among the
people of the United States; whose energy, ability,
and single-mindedness, in the prosecution of an arduous
task, have won for him the sympathy and support
of many of us in 'the old country.' I allude to
Professor Youmans. Quite as rapidly as in England,
the aim of these works was understood and appreciated
in the United States, and they brought me from this
side of the Atlantic innumerable evidences of good-
will. Year after year invitations reached me[2] to
visit America, and last year I was honoured with a
request so cordial, and signed by five-and-twenty
names so distinguished in science, in literature, and
in administrative position, that I at once resolved to
respond to it by braving not only the disquieting
oscillations of the Atlantic, but the far more disquieting ordeal of appearing in person before the people of
the United States.

This invitation, conveyed to me by my accomplished friend Professor Lesley, of Philadelphia, and
preceded by a letter of the same purport from your
scientific Nestor, the celebrated Joseph Henry, of
Washington, desired that I would lecture in some of
the principal cities of the Union. This I agreed to

---

[1] At whose hands it gives me pleasure to state I have always experienced honourable and liberal treatment.

[2] One of the earliest of these came from Mr. John Amory Lowell, of
Boston.

do, though much in the dark as to a suitable subject. In answer to my inquiries, however, I was given to understand that a course of lectures showing the uses of experiment in the cultivation of Natural Knowledge would materially promote scientific education in this country. And though such lectures involved the selection of weighty and delicate instruments, and their transfer from place to place, I at once resolved to meet the wishes of my friends as far as the time and means at my disposal would allow.

Experiments have two great uses—a use in discovery and verification, and a use in tuition. They were long ago defined as the investigator's language addressed to Nature, and to which she sends intelligible replies. These replies, however, usually reach the questioner in whispers too feeble for the public ear. But after the discoverer comes the teacher, whose function it is so to exalt and modify the experiments of his predecessor as to render them fit for public presentation. This secondary function I shall endeavour, in the present instance, to fulfil.

I propose to take a single department of natural philosophy, and illustrate, by means of it, the growth of scientific knowledge under the guidance of experiment. I wish, in this first lecture, to make you acquainted with certain elementary phenomena; then to point out to you how those theoretic principles by which phenomena are explained, take root, and flourish in the human mind, and afterwards to apply these principles to the whole body of knowledge covered by the lectures. The science of optics lends itself to this mode of treatment, and on it, therefore, I propose to

draw for the materials of the present course. It will
be best to begin with the few simple facts regarding
light which were known to the ancients, and to pass
from them in historic gradation to the more abstruse
discoveries of modern times.

All our notions of Nature, however exalted or however grotesque, have some foundation in experience.
The notion of personal volition in Nature had this basis.
In the fury and the serenity of natural phenomena the
savage saw the transcript of his own varying moods,
and he accordingly ascribed these phenomena to beings
of like passions with himself, but vastly transcending
him in power. Thus the notion of *causality*—the assumption that natural things did not come of themselves, but had unseen antecedents—lay at the root of
even the savage's interpretation of Nature. Out of
this bias of the human mind to seek for the antecedents of phenomena all science has sprung.

We will not now go back to man's first intellectual
gropings; much less shall we enter upon the thorny discussion as to how the groping man arose. We will take
him at a certain stage of his development, when, by evolution or sudden endowment, he became possessed of the
apparatus of thought and the power of using it. For
a time—and that historically a long one—he was limited
to mere observation, accepting what Nature offered,
and confining intellectual action to it alone. The apparent motions of sun and stars first drew towards them
the questionings of the intellect, and accordingly astronomy was the first science developed. Slowly, and with
difficulty, the notion of natural forces took root in the
human mind. No such notion, I may remark, is spon-
taneously generated, and the seedling of this one was the

actual observation of electric and magnetic attractions and repulsions. Slowly, and with difficulty, the science of mechanics had to grow out of this notion ; and slowly at last came the full application of mechanical principles to the motions of the heavenly bodies. We trace the progress of astronomy through Hipparchus and Ptolemy ; and, after a long halt, through Copernicus, Galileo, Tycho Brahe, and Kepler ; while from the high table-land of thought raised by these men Newton shoots upward like a peak, overlooking all others from his dominant elevation.

But other objects than the motions of the stars attracted the attention of the ancient world. Light was a familiar phenomenon, and from the earliest times we find men's minds busy with the attempt to render some account of it. But without experiment, which belongs to a later stage of scientific development, little progress could be made in this subject. The ancients, accordingly, were far less successful in dealing with light than in dealing with solar and stellar motions. Still they did make some progress. They satisfied themselves that light moved in straight lines ; they knew also that light was reflected from polished surfaces, and that the angle of incidence of the *rays* of light was equal to the angle of reflection. These two results of ancient scientific curiosity constitute the starting-point of our present course of lectures.

But in the first place it will be useful to say a few words regarding the source of light to be employed in our experiments. The rusting of iron is, to all intents and purposes, the slow burning of iron. It develops heat, and, if the heat be preserved, a high temperature

may be thus attained. The destruction of the first
Atlantic cable was probably due to heat developed in
this way. Other metals are still more combustible
than iron. You may light strips of zinc in a candle
flame, and cause them to burn almost like strips of
paper. But we must now expand our definition of
combustion, including under this term not only combustion in air, but also combustion in liquids. Water,
for example, contains a store of oxygen, which may
unite with and consume a metal immersed in it ; it is
from this kind of combustion that we are to derive the
heat and light employed in our present course.

The generation of this light and of this heat merits
a moment's attention. Before you is an instrument—
a small voltaic battery—in which zinc is immersed in
a suitable liquid. An attractive force is at this
moment exerted between the metal and the oxygen
of the liquid, actual union, however, being in the
first instance avoided. Uniting the two ends of the
battery by a thick wire, the attraction is satisfied,
the oxygen unites with the metal, zinc is consumed,
and heat, as usual, is the result of the combustion. A
power which, for want of a better name, we call an
electric current, passes at the same time through the
wire.

Cutting the thick wire in two, let the severed ends
be united by a thin one. It glows with a white heat.
Whence comes that heat ? The question is well worthy
of an answer. Suppose in the first instance, when
the thick wire is employed, that we permit the action
to continue until 100 grains of zinc are consumed, the
amount of heat generated in the battery would be
capable of accurate numerical expression. Let the

action then continue, with the thin wire glowing, until
100 grains of zinc are consumed.   Will the amount of
heat generated in the battery be the same as before ?
No, it will be less by the precise amount generated in
the thin wire outside the battery.   In fact, by adding
the internal heat to the external, we obtain for the
combustion of 100 grains of zinc a total which never
varies. We have here a beautiful example of that law of
constancy as regards natural energies, the establishment
of which is the greatest achievement of modern scientific

Fɪɢ. 1.

philosophy. By this arrangement, then, we are able to
burn our zinc at one place, and to exhibit the effects
of its combustion at a distant place.   In New York, for
example, we may have our grate and fuel ; but the heat
and light of our fire may be made to appear at San
Francisco.

Removing the thin wire and attaching to the severed
ends of the thick one two rods of coke, we obtain on
bringing the rods together (as in fig. 1), a small star of
light.   Now, the light to be employed in our lectures

is a simple exaggeration of this star. Instead of being produced by ten cells, it is produced by fifty. Placed in a suitable camera, provided with a suitable lens, this powerful source will give us all the light necessary for our experiments.

And here, in passing, I am reminded of the common delusion that the works of Nature, the human eye included, are theoretically perfect. The eye has grown for ages *towards* perfection; but ages of per-fect*ing* may be still before it. Looking at the dazzling light from our large battery, I see a luminous globe, but entirely fail to see the shape of the coke-points whence the light issues. The cause may be thus made clear: On the screen before you is projected an image of the carbon points, the *whole* of the lens in front of the camera being employed to form the image. It is not sharp, but surrounded by a halo which nearly obliterates the carbons. This arises from an imperfec-tion of the lens, called its *spherical aberration,* due to the fact that the circumferential and central rays have not the same focus. The human eye labours under a similar defect, and from this and other causes it arises that when the naked light from fifty cells is looked at, the blur of light upon the retina is sufficient to destroy the definition of the retinal image of the carbons. A long list of indictments might indeed be brought against the eye—its opacity, its want of symmetry, its lack of achromatism, its absolute blindness, in part. All these taken together caused Helmholtz to say that, if any optician sent him an instrument so full of defects, he would be justified in sending it back with the severest censure. But the eye is not to be judged from the stand-point of theory. It is not perfect, as I have said, but on its way to perfection. As a practical instrument, and taking the adjustments by which its defects are neutralized into account, it must ever remain a marvel to the reflecting mind.

The ancients, then, were aware of the rectilineal propagation of light. They knew that an opaque body, placed between the eye and a point of light, intercepted the light of the point. Possibly the terms ' ray ' and ' beam ' may have been suggested by those straight spokes of light which, in certain states of the atmosphere, dart from the sun at his rising and his setting. The rectilineal propagation of light may be illustrated at home by permitting the solar light to enter by a small aperture in a window shutter a dark room in which a little smoke has been diffused. In pure *air* you cannot see the beam, but in smoke you can, because the light, which passes unseen through the air, is scattered and revealed by the smoke particles, among which the beam pursues a straight course.

Or proceed in this way : Make a small hole in a closed window-shutter, before which stands a house or a tree, and place within the darkened room a white screen at some distance from the orifice. Every straight ray proceeding from the house or tree stamps its colour upon the screen, and the sum of all the rays will, therefore, be an image of the object. But, as the rays cross each other at the orifice, the image is inverted. At present we may illustrate and expand the subject thus : In front of our camera is a large opening (L, fig. 2), from which the lens has been removed, and which is closed at present by a sheet of tinfoil. Pricking by means of a common sewing-needle a small aperture in the tin-foil, an inverted image of the carbon-points starts forth upon the screen. A dozen

apertures will give a dozen images, a hundred a
hundred, a thousand a thousand. But, as the apertures
come closer to each other, that is to say, as the tin-foil
between the apertures vanishes, the images overlap
more and more. Removing the tin-foil altogether, the
screen becomes uniformly illuminated. Hence the
light upon the screen may be regarded as the overlap-
ping of innumerable images of the carbon-points. In

Fig. 2.

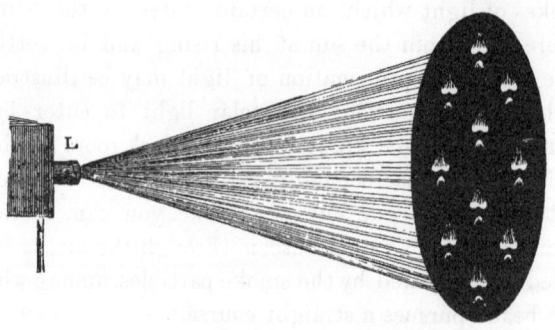

like manner the light upon every white wall on a
cloudless day may be regarded as produced by the
superposition of innumerable images of the sun.

The law that the angle of incidence is equal to the
angle of reflection has a bearing upon a theory, to be
subsequently mentioned, which renders its simple illustration here desirable. A straight lath (pointing
to 5 in fig. 3) is placed as an index perpendicular
to a small looking-glass (M) capable of rotation.
A beam of light is first received upon the glass and reflected back along the line of its incidence. Though
the incident and the reflected beams pass in opposite
directions, they do not jostle or displace each other.

The index being turned, the mirror turns along with it,
and at each side of the index the incident and the reflected beams (L o, o R), are seen tracking themselves
through the dust of the room. The mere inspection of
the two angles enclosed between the index and the two
beams suffices to show their equality. Small elastic balls
impinging on the looking-glass would follow the course

Fig. 3.

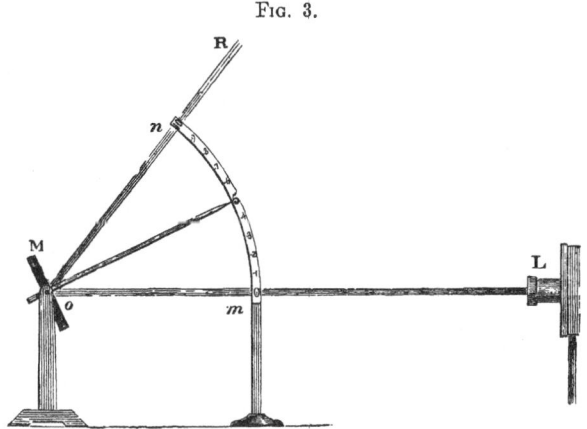

of the reflected light. A card placed edgeways upon a
table without inclination to the right or to the left
is said to be perpendicular to the plane of the table.
In the case of light the incident and reflection rays
always lie in a plane perpendicular to the reflecting
surface.

This simple apparatus enables us to illustrate another
law of great practical importance, namely, that, when a
mirror rotates, the angular velocity of a beam reflected
from it is twice that of the reflecting mirror. A simple
experiment will make this plain to you. The arc (m n fig.
3) before you is divided into ten equal parts, and when

the incident beam and the index cross the zero of the
graduation, both the incident and reflected beams are
horizontal. Moving the index of the mirror to 1,
the reflected beam cuts the arc at 2 ; moving the index
to 2, the arc is cut at 4 ; moving the index to 3, the arc
is cut at 6 ; moving the index to 4, the arc is cut at 8 ;
finally, moving the index to 5, the arc is cut at 10 (as
in the figure). In every case the reflected beam moves
through twice the angle of the mirror.

One of the problems of science, on which scientific
progress mainly depends, is to help the senses of man
by carrying them into regions which could never be
attained without such help. Thus we arm the eye with
the telescope when we want to sound the depths of
space, and with the microscope when we want to explore motion and structure in their infinitesimal dimensions. Now, this law of angular reflection, coupled
with the fact that a beam of light possesses no weight,
gives us the means of magnifying small motions to an
extraordinary degree. Thus, by attaching mirrors to
his suspended magnets, and by watching the images of
divided scales reflected from the mirrors, the celebrated
Gauss was able to detect the slightest thrill of variation
on the part of the earth's magnetic force. By a similar
arrangement the feeble attractions and repulsions of
the diamagnetic force have been made manifest. The
minute elongation of a bar of metal by the mere warmth
of the hand may be so magnified by this method as to
cause the index-beam to move from the ceiling to the
floor of this room. The lengthening of a bar of iron
when it is magnetized may be also thus demonstrated.
Helmholtz long ago employed this method to render
evident to his students the classical experiments of

Du Bois Raymond on animal electricity; while in Sir William Thomson's reflecting galvanometer the principle receives one of its latest applications.

For more than a thousand years no step was taken in optics beyond this law of reflection. The men of the Middle Ages, in fact, endeavoured on the one hand to develope the laws of the universe *à priori* out of their own consciousness, while many of them were so occupied with the concerns of a future world that they looked with a lofty scorn on all things pertaining to this one. Speaking of the natural philosophers of his time, Eusebius says, 'It is not through ignorance of the things admired by them, but through contempt of their useless labour, that we think little of these matters, turning our souls to the exercise of better things.' So also Lactantius—'To search for the causes of things; to inquire whether the sun be as large as he seems; whether the moon is convex or concave; whether the stars are fixed in the sky, or float freely in the air; of what size and of what material are the heavens; whether they be at rest or in motion; what is the magnitude of the earth; on what foundations is it suspended or balanced;—to dispute and conjecture upon such matters is just as if we chose to discuss what we think of a city in a remote country, of which we never heard but the name.'[1]

---

[1] The spirit of those ancient heroes of the faith is still to be found in unexpected places. In the April number of the *Contemporary Review*, after describing how modern science came to be what it is, my friend Dr. Acland puts the following language into the mouth of Bishop Wilson:—'What is surprising to me is the labour that you have taken to attain so very little. You deserve for this the utmost credit a reasonable being can desire; for you, being so accurate and so painstaking.

As regards the refraction of light, the course of real inquiry was resumed in 1100 by an Arabian philosopher named Alhazen. Then it was taken up in succession by Roger Bacon, Vitellio, and Kepler. One of the most important occupations of science is the determination, by precise measurements, of the quantitative relations of phenomena ; the value of such measurements depending greatly upon the skill and conscientiousness of the man who makes them. Vitellio appears to have been both skilful and conscientious, while Kepler's habit was to rummage through the observations of his predecessors, to look at them in all lights, and thus distil from them the principles which united them. He had done this with the astronomical measurements of Tycho Brahe, and had extracted from them the celebrated 'laws of Kepler.' He did it also with Vitellio's measurements of refraction. But in this case he was not successful. The principle, though a simple one, escaped him, and it was first discovered by Willebrod Snell, about the year 1621.

Less with the view of dwelling upon the phenomenon itself than of introducing it in a form which will render intelligible to you subsequently the play of theoretic thought in Newton's mind, the fact of refraction may be here demonstrated. I will not do this by drawing

seem well aware of the uncertainty of some of your data, and of the possible futility, therefore, of some of your conclusions. For I am told that with all your pains, your sciences contain within them so many examples of proved errors, that, being candid men, you must often feel the material ground under your feet to be very slippery.' Schelling thus expresses his contempt for experimental knowledge : 'Newton's Optics is the greatest illustration of a whole structure of fallacies, which in all its parts is founded on observation and experiment.' There are some small imitators of Schelling still in Germany.

the course of the beam with chalk on a black board, but by causing it to mark its own white track before you. A shallow circular vessel (R I G, fig. 4), with a glass face, half filled with water rendered barely turbid by the admixture of a little milk or the precipitation of a little mastic, is placed upon its edge with its glass face vertical. By means of a small plane reflector (M), and through a slit (I) in the hoop surrounding the vessel, a

Fig. 4.

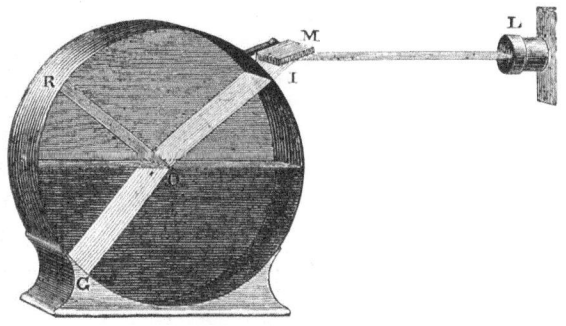

beam of light is admitted in any required direction. It impinges upon the water (at O), enters it, and tracks itself through the liquid in a sharp, bright band (O G). Meanwhile the beam passes unseen through the air above the water, for the air is not competent to scatter the light. A puff of tobacco-smoke into this space at once reveals the track of the incident-beam. If the incidence be vertical, the beam is unrefracted. If oblique, its refraction at the common surface of air and water (at O), is rendered clearly visible. It is also seen that *reflection* (along O R) accompanies refraction, the beam dividing itself at the point of incidence into a refracted and a reflected portion.

The law by which Snell connected together all the measurements executed up to his time, is this: Let A B C D (Fig 5.) represent the outline of our circular vessel, A C being the water-line. When the beam is incident along B E, which is perpendicular to A C, there is no refraction. When it is incident along *m* E, there is refraction: it is bent at E and strikes the circle at *n*. When it is incident along *m′* E, there is also refraction at E, the beam striking the point *n′*.

Fig. 5.

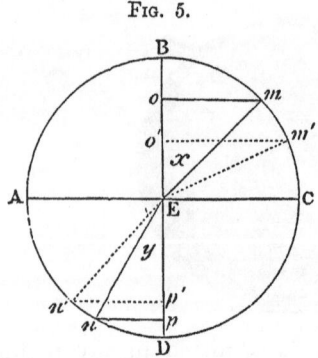

From the ends of the incident beams, let the perpendiculars *m o*, *m′ o′* be drawn upon B D, and from the ends of the refracted beams let the perpendiculars *p n*, *p′ n′* be also drawn. Measure the lengths of *o m* and of *p n*, and divide the one by the other. You obtain a certain quotient. In like manner divide *m′ o′* by the corresponding perpendicular *p′ n′*; you obtain in each case the same quotient. Snell, in fact, found this quotient to be *a constant quality* for each particular substance, though it varied in amount from substance to substance. He called the quotient the *index of refraction*.

In all cases where the light is incident from air upon the surface of a solid or a liquid, or, more generally still, when the incidence is from a less highly refracting to a more highly refracting medium, the reflection is *partial*. In this case the most powerfully reflecting substances either transmit or absorb a portion of the incident light. At a perpendicular incidence water reflects only 18 rays out of every 1,000; glass reflects only 25 rays, while mercury reflects 666. When the rays strike the surface obliquely the reflection is augmented. At an incidence of 40°, for example, water reflects 22 rays, at 60° it reflects 65 rays; at 80° 333 rays; while at an incidence of 89½°, where the light almost grazes the surface, it reflects 721 rays out of every 1,000. Thus, as the obliquity increases, the reflection from water approaches, and finally quite overtakes, the reflection from mercury; but at no incidence, however great, is the reflection from water, mercury, or any other substance, *total*.

Still, total reflection may occur, and with a view to understanding its subsequent application in the Nicol's prism, it is necessary to state when it occurs. This leads me to the enunciation of a principle which underlies all optical phenomena—the principle of reversibility.[1] In the case of refraction, for instance, when the ray passes obliquely from air into water, it is bent *towards* the perpendicular; when it passes from water to air, it is bent *from* the perpendicular, and accurately reverses its course. Thus in fig. 5, if *m* ᴇ *n* be the track taken by a ray in passing from air into water, *n* ᴇ *m*

---

[1] From this principle Sir John Herschel deduces in a simple and elegant manner the fundamental law of reflection.—See *Familiar Lectures*, p. 236.

will be its track in passing from water into air. Let us push this principle to its consequences. Supposing the light, instead of being incident along *m* E or *m'* E' were incident as close as possible along C E, (fig. 6); suppose, in other words, that it just grazes the surface before entering the water. After refraction it will pursue the course E *n''*. Conversely, if the light start from *n''*, and be incident at E, it will on escaping into the air just graze the surface of the water. The question now arises, what will occur supposing the ray from the water follows the course *n'''* E, which lies beyond *n''* E?

Fig. 6.

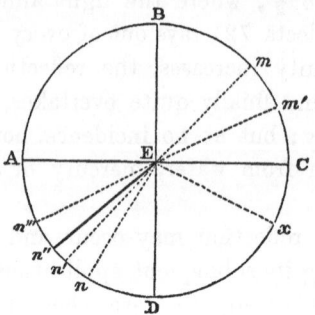

The answer is, it will not quit the water at all, but will be *totally* reflected (along E *x*). At the under surface of the water, moreover, the law is just the same as at its upper surface, the angle of incidence (D E *n'''*) being here also equal to the angle of reflection (D E *x*).

Total reflection may be thus simply illustrated :— Place a shilling in a drinking-glass, and tilt the glass so that the light from the shilling shall fall with the necessary obliquity upon the water surface above it. Look upwards towards that surface, and you see the image of the shilling shining there as brightly as the shilling itself. Thrust the closed end of a glass test-tube into water, and incline the tube. When the inclination is sufficient, the horizontal light falling upon the tube cannot enter the air within it, but is totally reflected upward; when looked down upon, such a tube looks quite as bright as burnished silver. Pour a little water into the tube; as the liquid rises total reflection is abolished, and with it the lustre, leaving a gradually diminishing shining zone, which disappears wholly when the level of the water within the tube reaches that without it. Any glass tube, with its end stopped water-tight, will produce this effect, which is both beautiful and instructive.

Total reflection never occurs except in the attempted passage of a ray from a more refracting to a less refracting medium; but in this case, when the obliquity is sufficient, it always occurs. The mirage of the desert and other phantasmal appearances in the atmosphere are in part due to it. When, for example, the sun heats an expanse of sand, the layer of air in contact with the sand becomes lighter and less refracting than the air above it; consequently, the rays from a distant object striking very obliquely on the surface of the heated stratum, are sometimes totally reflected upwards, thus producing images similar to those produced by water. I have seen the image of a rock called Mont Tombeline inverted on the heated strand of Normandy near Avranches; and by such delusive appearances the thirsty soldiers of the French army in Egypt were for a time greatly tantalized.

The angle which marks the limit beyond which total reflection takes place is called the *limiting angle* (it is marked in fig. 6 by the strong line E n''). It must evidently diminish as the refractive index increases. For water it is 48½°, for flint glass 38° 41′, and for diamond 23° 42′. Thus all the light incident from two complete quadrants, or 180°, in the case of diamond, is condensed into an angular space of 47° 22′ (twice 23° 42′) by refraction. Coupled with its great refraction, are the great dispersive and great reflective powers of diamond ; hence the extraordinary radiance of the gem, both as regards white light and prismatic light.[1]

In 1676 an impulse was given to optics by astronomy. In that year Olav Rœmer, a learned Dane, was engaged at the Observatory of Paris in observing the eclipses of Jupiter's moons. The planet, whose distance from the sun is 475,693,000 miles, has four satellites. We are now only concerned with the one nearest to the planet. Rœmer watched this moon, saw it move round in front of the planet, pass to the other side of it, and then plunge into Jupiter's shadow, behaving like a lamp suddenly extinguished : at the second edge of the shadow he saw it reappear like a lamp suddenly lighted. The moon thus acted the part of a signal light to the astronomer, and enabled him to tell exactly its time of revolution. The period between two successive lightings up of the lunar lamp he found to be 42 hours, 28 minutes, and 35 seconds.

This measurement of time was so accurate, that

---

[1] With regard to the total reflection of light within a jet of water, where the jet shines like molten lava as it falls, it might well be asked, if the reflection be *total* and *internal*, how is the jet seen? Were the water *pure*, I believe the jet could not be seen ; the light, as Emerson says, would in this case hide itself in transparency. The jet is seen as our beam is seen in the experiment on refraction (Fig. 4)— through the scattering of the light by mechanically suspended particles.

having determined the moment when the moon emerged from the shadow, the moment of its hundredth appearance could also be determined. In fact it would be 100 times 42 hours, 28 minutes, 35 seconds, after the first observation.

Rœmer's first observation was made when the earth was in the part of its orbit nearest Jupiter. About six months afterwards, the earth being then at the opposite·side of its orbit, when the little moon ought to have made its hundredth appearance, it was found unpunctual, being fully 15 minutes behind its calculated time. Its appearance, moreover, had been growing gradually later, as the earth retreated towards the part of its orbit most distant from Jupiter. Rœmer reasoned thus:—' Had I been able to remain at the other side of the earth's orbit, the moon might have appeared always at the proper instant; an observer placed there would probably have seen the moon 15 minutes ago, the retardation in my case being due to the fact that the light requires 15 minutes to travel from the place where my first observation was made to my present position.'

This flash of genius was immediately succeeded by another. ' If this surmise be correct,' Rœmer reasoned, ' then as I approach Jupiter along the other side of the earth's orbit, the retardation ought to become gradually less, and when I reach the place of my first observation, there ought to be no retardation at all.' He found this to be the case, and thus not only proved that light required time to pass through space, but also determined its rate of propagation.

The velocity of light, as determined by Rœmer, is 192,500 miles in a second.

For a time, however, the observations and reasonings of Rœmer failed to produce conviction. They were doubted by Cassini, Fontenelle, and Hooke. Subsequently came the unexpected corroboration of Rœmer by the English astronomer, Bradley, who noticed that the fixed stars did not really appear to be fixed, but that they describe little orbits in the heavens every year. The result perplexed him, but Bradley had a mind open to suggestion, and capable of seeing, in the smallest fact, a picture of the largest. He was one day upon the Thames in a boat, and noticed that as long as his course remained unchanged, the vane upon his masthead showed the wind to be blowing constantly in the same direction, but that the wind appeared to vary with every change in the direction of his boat. 'Here,' as Whewell says, 'was the image of his case. The boat was the earth, moving in its orbit, and the wind was the light of a star.'

We may ask in passing, what without the faculty which formed the 'image,' would Bradley's wind and vane have been to him? A wind and vane, and nothing more. You will immediately understand the meaning of Bradley's discovery. Imagine yourself in a motionless railway-train with a shower of rain descending vertically downwards. The moment the train begins to move the rain-drops begin to slant, and the quicker the motion of the train the greater is the obliquity. In a precisely similar manner the rays from a star vertically overhead are caused to slant by the motion of the earth through space. Knowing the speed of the train, and the obliquity of the falling rain, the velocity of the drops may be calculated; and knowing the speed of the earth in her orbit, and the obliquity of the rays due to this cause, we can calculate just as easily the velocity of light. Bradley did this, and the 'aberration of light,' as his discovery is called, enabled him to assign to it a velocity almost identical with that deduced by Rœmer from a totally different method of observation. Subsequently Fizeau, employing not planetary or stellar distances, but simply the breadth of the city of Paris, determined the velocity of light: while after him Foucault—a man of the rarest mechanical genius—solved the problem without quitting his private room. Owing to an error in the determination of the earth's distance from the sun, the velocity assigned to light by both Rœmer and Bradley is too great. With a close approximation to accuracy it may be regarded as 186,000 miles a second.

By Rœmer's discovery, the notion entertained by Descartes, and espoused by Hooke, that light is propagated instantly through space, was overthrown. But the establishment of its velocity through stellar space led to speculations regarding its velocity in transparent terrestrial substances. The index of refraction of a ray passing from air into water is $\frac{4}{3}$. Newton assumed these numbers to mean that the velocity of light in water being 4 its velocity in air is 3; and he deduced the phenomena of refraction from this assumption. The reverse has since been proved to be the case—that is to say, the velocity of light in water being 3, its velocity in air is 4; but both in Newton's time and ours the same great principle determined, and determines, the course of light in all cases. In passing from point to point, whatever be the media in its path, or however it may be reflected, light takes the course which occupies *least time*. Thus in fig. 4, taking its velocity in air

and in water into account, the light reaches G from I
more rapidly by travelling first to O, and there changing
its course, than if it proceeded straight from I to G.
This is readily comprehended, because in the latter case
it would pursue a greater distance through the water,
which is the more retarding medium.

Snell's law of refraction is one of the corner-stones of
optical science, and its applications to-day are million-
fold. Immediately after its discovery Descartes applied it
to the explanation of the rainbow. A beam of solar light
falling obliquely upon a rain-drop is refracted on entering the drop. It is in part reflected at the back of the
drop, and on emerging it is again refracted. By these
two refractions, at its entrance and at its emergence, the
beam of light is decomposed, quitting the drop resolved
into its coloured constituents. The light thus reaches
the eye of an observer facing the drop, and with his
back to the sun.

Conceive a line drawn from the sun to the observer's
eye and prolonged beyond the observer. Conceive
another line drawn, enclosing an angle of $42\frac{1}{2}°$ with the
line drawn from the sun, and prolonged to the falling
shower. Along this second line the rain-drop, at its
remote end, when struck by a sunbeam, will send a ray
of red light. Every other drop similarly situated, that
is, every drop at an angular distance of $42\frac{1}{2}°$ from the
line aforesaid, will do the same. A circular band of red
light is thus formed, which may be regarded as a portion
of the base of a cone, having the rays which form
the band for an envelope, and its apex at the observer's
eye. Because of the magnitude of the sun, the angular
width of this red band will be half a degree.

From the eye of the observer conceive another line

to be drawn, enclosing an angle, not of $42\frac{1}{2}°$, but of $40\frac{1}{2}°$, with the line drawn from the eye to the sun. Along this line a solar beam striking a rain-drop will send violet light to the eye. All drops at the same angular distance will do the same, and we shall therefore obtain a band of violet light of the same width as the red band. These two bands constitute the limiting colours of the rainbow, and between them the bands corresponding to the other colours lie.

Thus the line drawn from the observer to the *middle* of the bow and the line drawn through the observer to the sun always enclose an angle of about 41°; to account for this was the great difficulty, which remained unsolved up to the time of Descartes.

Taking a pen in hand and calculating by means of Snell's law the track of every ray through a rain-drop, Descartes found that, at one particular angle, the rays emerged from the drop *almost parallel to each other*. They were thus enabled to preserve their intensity through long atmospheric distances. At all other angles the rays quitted the drop *divergent*, and through this divergence became so enfeebled as to be practically lost to the eye. The angle of parallelism here referred to was that of forty-one degrees, which observation had proved to be invariably associated with the rainbow.

And here we may devote a moment to a question which has often been the subject of public discussion— whether, namely, a rainbow which spans a tranquil sheet of water is ever seen reflected in the water? Supposing you cut an arch out of pasteboard, of the apparent width of the rainbow, and paint upon it the colours of the bow; such a painted arch, spanning still water, would, if not too distant, undoubtedly be seen reflected in the water. The coloured rays from such an arch would be emitted in all directions, those striking the water at the proper angle, and reflected to the eye, giving the image of the arch. But the rays effective in the rainbow are emitted only in the direction fixed by the angle of 41°. Those rays, therefore, which are scattered from the drops upon the water, do not carry along with them the necessary condition of parallelism; and, hence, though the cloud on which the bow is painted may be reflected from the water, we can have no reflection of the bow itself.

In the rainbow a new phenomenon was introduced —the phenomenon of colour. And here we arrive at one of those points in the history of science, when great men's labours so intermingle that it is difficult to assign to each worker his precise meed of honour. Descartes was at the threshold of the discovery of the composition of solar light; but for Newton was reserved the enunciation of the true law. He went to work in this way: Through the closed window-shutter of a room he pierced an orifice, and allowed a thin sunbeam to pass through it. The beam stamped a round white image of the sun on the opposite wall of the room. In the path of this beam Newton placed a prism, expecting to see the beam refracted, but also expecting to see the image of the sun, after refraction, round. To his astonishment, it was drawn out to an image with a length five times its breadth. It was, moreover, no longer white, but divided into bands of different colours. Newton saw immediately that solar light was *composite*, not simple. His elongated image revealed to him the fact that some constituents of the

light were more deflected by the prism than others,
and he concluded, therefore, that white solar light was
a mixture of lights of different colours and of different
degrees of refrangibility.

Let us reproduce this celebrated experiment. On
the screen is now stamped a luminous disk, which may
stand for Newton's image of the sun. Causing the beam
(from L, fig. 7) which produces the disk to pass through
a lens (E) which forms an image of the aperture, and

FIG. 7.

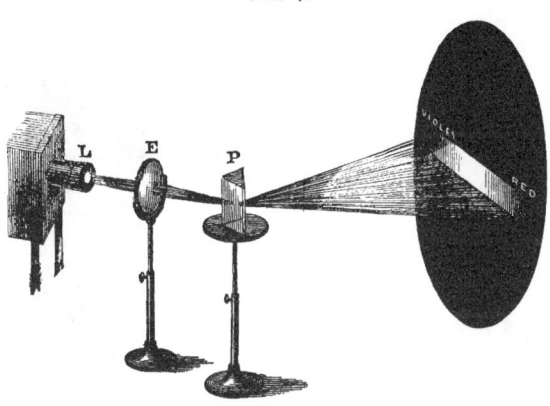

then through a prism (P), we obtain Newton's coloured
image, with its red and violet ends, which he called *a
spectrum*. Newton divided the spectrum into seven
parts—red, orange, yellow, green, blue, indigo, violet;
which are commonly called the seven primary or prismatic colours. The drawing out of the white light
into its constituent colours is called *dispersion*.

This was the first *analysis* of solar light by Newton;
but the scientific mind is fond of verification, and never
neglects it where it is possible. Newton completed his

proof by *synthesis* in this way: The spectrum now
before you is produced by a glass prism. Causing the
decomposed beam to pass through a second similar
prism, but so placed that the colours are refracted back
and reblended, the perfectly white luminous disk is
restored.

In this case, refraction and dispersion are simultaneously abolished. Are they always so? Can we have
the one without the other? It was Newton's conclusion that we could not. Here he erred, and his error,

FIG. 8.

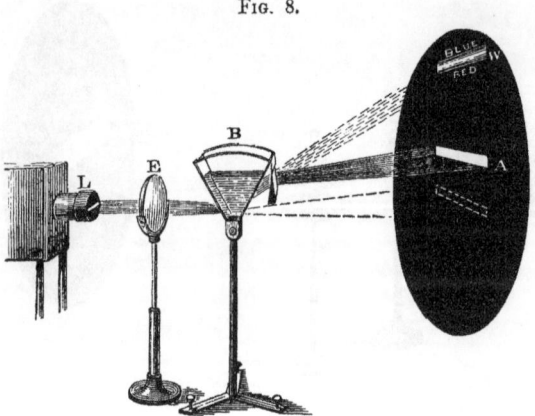

which he maintained to the end of his life, retarded
the progress of optical discovery. Dolland subsequently proved that, by combining two different kinds
of glass, the colours can be extinguished, still leaving
a residue of refraction, and he employed this residue
in the construction of achromatic lenses—lenses
yielding no colour—which Newton thought an impossi-
bility. By setting a water-prism—water contained in
a wedge-shaped vessel with glass sides (B, fig. 8)—in
opposition to a wedge of glass (to the right of B), this

point can be illustrated before you. We have first of all the position (dotted) of the unrefracted beam marked upon the screen; then we produce the narrow water-spectrum (W); finally, by introducing a flint-glass prism, we refract the beam back, until the colour disappears (at A). The image of the slit is now *white;* but you see that, though the dispersion is abolished, there remains a very sensible amount of refraction.

This is the place to illustrate another point bearing upon the instrumental means employed in these lectures. Bodies differ widely from each other as to their powers of refraction and dispersion. Note the position of the water-spectrum upon the screen. Altering in no particular the wedge-shaped vessel, but simply substituting for the water the transparent bisulphide of carbon, you notice how much higher the beam is thrown, and how much richer is the display of colour. To augment the size of our spectrum we here employ (at L) a slit instead of a circular aperture.[1]

The synthesis of white light may be effected in three ways, which are now worthy of attention: Here, in the first instance, we have a rich spectrum

---

[1] The low dispersive power of water masks, as Helmholtz has remarked, the imperfect achromatism of the eye. With the naked eye, I can see a distant blue disk sharply defined, but not a red one. I can also see the lines which mark the upper and lower boundaries of a horizontally refracted spectrum sharp at the blue end, but ill-defined at the red end. Projecting a luminous disk upon a screen, and covering one semicircle of the aperture with a red and the other with a blue or green glass, the difference between the apparent sizes of the two semicircles is in my case, and in numerous other cases, extraordinary. Many persons, however, see the apparent sizes of the two semicircles reversed. If with a spectacle glass I correct the dispersion of the red light over the retina, then the blue ceases to give a sharply-defined image. Thus examined the departure of the eye from achromatism appears very gross indeed.

produced by the decomposition of the beam (from L
fig. 9). One face of the prism (P) is protected by a
diaphragm (not shown in the figure), with a longitudinal slit, through which the beam passes into the prism.
It emerges decomposed at the other side. I permit the
colours to pass through a cylindrical lens (C), which so
squeezes them together as to produce upon the screen a
sharply-defined rectangular image of the longitudinal
slit (now upright). In that image the colours are re-

FIG. 9.

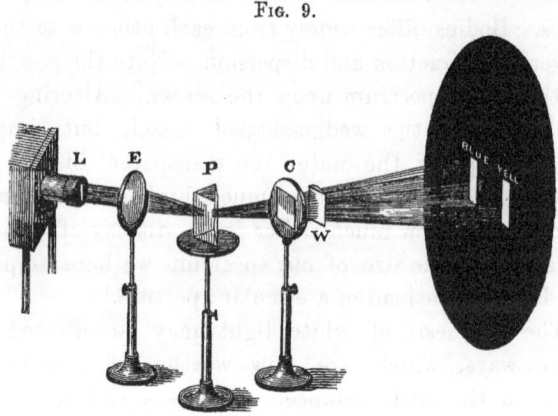

blended, and you see it perfectly white. Between the
prism and the cylindrical lens may be seen the colours
tracking themselves through the dust of the room.
Cutting off the more refrangible fringe by a card, the
rectangle is seen red; cutting off the less refrangible
fringe, the rectangle is seen blue. By means of a thin
glass prism (W), I deflect one portion of the colours, and
leave the residual portion. On the screen are now two
coloured rectangles produced in this way. These are
complementary colours—colours which, by their union,

produce white. Note that, by judicious management, one of these colours is rendered *yellow*, and the other *blue*. I withdraw the thin prism ; yellow and blue immediately commingle, and we have *white* as the result of their union. On our way, then, we remove the fallacy, first exposed by Helmholtz, that the mixture of blue and yellow lights produces green.

Restoring the circular aperture, we obtain once more a spectrum like that of Newton. By means of a lens, we gather up these colours, and build them together, not to an image of the aperture, but to an image of the carbon-points themselves.

Finally, in virtue of the persistence of impressions upon the retina, by means of a rotating disk, on which are spread in sectors the colours of the spectrum, we blend together the prismatic colours *in the eye itself*, and thus produce the impression of whiteness.

Having unravelled the interwoven constituents of white light, we have next to inquire, What part the constitution so revealed enables this agent to play in Nature ?   To it we owe all the phenomena of colour ; and yet not to it alone, for there must be a certain relationship between the ultimate particles of natural bodies and white light to enable them to extract from it the luxury of colour.   But the function of natural bodies is here *selective,* not *creative.*   There is no colour generated by any natural body whatever.   Natural bodies have showered upon them, in the white light of the sun, the sum total of all possible colours, and their action is limited to the sifting of that total, the appropriating from it of the colours which really belong to them and the rejecting of those which do not.   It will fix this subject in your minds if I say that it is the portion

of light which they reject, and not that which belongs
to them, that gives bodies their colours.

Let us begin our experimental inquiries here by
asking, What is the meaning of blackness? Pass a
black ribbon through the colours of the spectrum; it
quenches all of them. The meaning of blackness is
thus revealed—it is the result of the absorption of *all*
the constituents of solar light. Pass a red ribbon
through the spectrum. In the red light the ribbon is a
vivid red. Why? Because the light that enters the
ribbon is not quenched or absorbed, but in great part
sent back to the eye. Place the same ribbon in the green
of the spectrum; it is black as jet. It absorbs the green
light, and leaves the space on which it falls a space of
intense darkness. Place a green ribbon in the green
of the spectrum. It shines vividly with its proper
colour; transfer it to the red, it is black as jet. Here
it absorbs all the light that falls upon it, and offers
mere darkness to the eye.

Thus, when white light is employed, the red sifts
it by quenching the green, and the green sifts it
by quenching the red, both exhibiting the residual
colour. The process through which natural bodies
acquire their colours is therefore a *negative* one.
The colours are produced by subtraction, not by addi-
tion. This red glass is red because it destroys all the
more refrangible rays of the spectrum. This blue
liquid is blue because it destroys all the less refrangible
rays. Both together are opaque because the light
transmitted by the one is quenched by the other. In
this way, by the union of two transparent substances
we obtain a combination as dark as pitch to solar light.
This other liquid, finally, is purple because it destroys

the green and the yellow, and allows the terminal
colours of the spectrum to pass unimpeded. From the
blending of the blue and the red this gorgeous purple
is produced.

One step further for the sake of exactness. The light
which falls upon a body is divided into two portions,
one of which is reflected from the surface of the
body; and this is of the same colour as the incident light.
If the incident light be white the superficially reflected
light will also be white. Solar light, for example, reflected from the surface of even a black body, is white.
The blackest camphine smoke in a dark room through
which a sunbeam passes from an aperture in the window-
shutter, renders the track of the beam white, by the
light scattered from the surfaces of the soot particles.
The moon appears to us as if

‘ Clothed in white samite, mystic, beautiful ; ’

but were she covered with the blackest velvet she would
still hang in the heavens as a white orb, shining upon
our world substantially as she does now.

The second portion of the light enters the body,
and upon its treatment there the colour of the body
depends. Let us analyse the action of pigments upon
light. They are composed of particles mixed with a
vehicle ; but how intimately soever the particles may
be blended, they still remain particles, separated it may
be by exceedingly minute distances, but still separated.
To use the scientific phrase, they are not optically
continuous. Now, wherever optical continuity is
ruptured we have reflection of the incident light. It
is the multitude of reflections at the limiting surfaces
of the particles that prevents light from passing

D

through glass, or rock-salt, when these transparent substances are pounded into powder. The light here is exhausted in a waste of echoes, not extinguished by true absorption. It is the same kind of reflecticn that renders the thunder-cloud so impervious to light. Such a cloud is composed of particles of water mixed with particles of air, both separately transparent, but practically opaque when thus mixed together.

In the case of pigments, then, the light is *reflected* at the limiting surfaces of the particles, but it is in part *absorbed* within the particles. The reflection is necessary to send the light back to the eye ; the absorption is necessary to give the body its colour. The same remarks apply to flowers. The rose is red in virtue, not of the light reflected from its surface, but of light which has entered its substance, which has been reflected from surfaces within, and which in returning *through* the substance has its green extinguished. A similar process in the case of hard green leaves extinguishes the red, and sends green light from the body of the leaves to the eye.

All bodies, even the most transparent, are more or less absorbent of light. Take the case of water: in small quantities it does not sensibly affect the light. A glass cell of clear water interposed in the track of our beam does not perceptibly change any one of the colours of the spectrum derived from the beam. Still absorption, though insensible, has here occurred, and to render it sensible we have only to increase the depth of the water through which the light passes. Instead of a cell an inch thick, let us take a layer, ten or fifteen feet thick : the colour of the water is then very evident. By augmenting the thickness we absorb more of the light, and by making the thickness very great we absorb the light altogether. Lampblack or pitch can do no more, and the only difference between them and water is that a very small depth in their case suffices to extinguish all the light. The difference between the highest known transparency, and the highest known opacity, is one of degree merely.

If, then, we render water sufficiently deep to quench all the light; and if from the interior of the water no light reaches the eye, we have the condition necessary to produce blackness. Looked properly down upon there are portions of the Atlantic Ocean to which one would hardly ascribe a trace of colour: at the most a tint of dark indigo reaches the eye. The water, in fact, is practically *black*, and this is an indication both of its depth and purity. But the case is entirely changed when the ocean contains solid particles in a state of mechanical suspension, capable of sending light back to the eye.

Throw, for example, a white pebble into the blackest Atlantic water; as it sinks it becomes greener and greener, and, before it disappears, it reaches a vivid blue green. Break such a pebble into fragments, these will behave like the unbroken mass: grind the pebble to powder, every particle will yield its modicum of green; and if the particles be so fine as to remain suspended in the water, the scattered light will be a uniform green. Hence the greenness of shoal water. You go to bed with the black water of the Atlantic around you. You rise in the morning and find it a vivid green; and you correctly infer that you are crossing the bank of Newfoundland. Such water is found charged with fine matter in a state of mechanical suspension. The light from the bottom may sometimes come into play, but it is not necessary. The subaqueous foam generated by the screw or paddle-wheels of a steamer also sends forth a vivid green. The foam here furnishes a *reflecting surface,* the water between the eye and it the *absorbing medium.*

Nothing can be more superb than the green of the Atlantic waves when the circumstances are favourable to the exhibition of the colour. As long as a wave remains unbroken no colour appears, but when the foam just doubles over the crest like an Alpine snow-cornice, under the cornice we often see a display of the most exquisite green. It is metallic in its brilliancy. But the foam is necessary to its production. The foam is first illuminated, and it scatters the light in all directions; the light which passes through the higher portion of the wave alone reaches the eye, and gives to that portion its matchless colour. The folding of the wave, producing as it does, a series of longitudinal protuberances and furrows which act like cylindrical lenses, introduces variations in the intensity of the light, and materially enhances its beauty.

We are now prepared for the further consideration of a point already adverted to, and regarding which error long found currency. You will find it stated in many books that blue and yellow lights mixed together produce green. But blue and yellow have been just proved to be complementary colours, producing white by their mixture. The mixture of blue and yellow *pigments* undoubtedly produces green, but the mixture of pigments is totally different from the mixture of lights.

Helmholtz, who first proved yellow and blue to be

complementary colours, has revealed the cause of the green in the case of pigments. No natural colour is *pure*. A blue liquid or a blue powder permits not only the blue to pass through it, but a portion of the adjacent green. A yellow powder is transparent not only to the yellow light, but also in part to the adjacent green. Now, when blue and yellow are mixed together, the blue cuts off the yellow, the orange, and the red; the yellow, on the other hand, cuts off the violet, the indigo, and the blue. Green is the only colour to which both are transparent, and the consequence is that, when white light falls upon a mixture of yellow and blue powders, the green alone is sent back to the eye. You have already seen that the fine blue ammonia-sulphate of copper transmits a large portion of green, while cutting off all the less refrangible light. A yellow solution of picric acid also allows the green to pass, but quenches all the more refrangible light. What must occur when we send a beam through both liquids? The experimental answer to this question is now before you: the green band of the spectrum alone remains upon the screen.

The impurity of natural colours is strikingly illustrated by an observation recently communicated to me by Mr. Woodbury. On looking through a blue glass at green leaves in sunshine, he saw the superficially reflected light blue. The light, on the contrary, which came from the body of the leaves was crimson. On examination, I found that the glass employed in this observation transmitted both ends of the spectrum, the red as well as the blue, and that it quenched the middle. This furnished an easy explanation of the effect. In the delicate spring

foliage the blue is for the most part absorbed, and a
light, mainly yellowish green, but containing a considerable quantity of red, escapes from the leaf to the
eye. On looking at such foliage through the violet
glass, the green and the yellow are stopped, and the red
alone reaches the eye. Thus regarded, therefore, the
leaves appear like faintly-blushing roses, and present a
very beautiful appearance. With the blue ammonia-
sulphate of copper, which transmits no red, this effect
is not obtained.

As the year advances the crimson gradually hardens
to a coppery red; and in the dark green leaves of old
ivy it is entirely absent. Permitting a concentrated
beam of white light to fall upon fresh leaves in a dark
room, the sudden change from green to red, and from
red back to green, when the violet glass is alternately
introduced and withdrawn, is very surprising. Looked
at through the same glass the meadows in May appear
of a warm purple. With a solution of permanganate
of potash, which, while it quenches the centre of the
spectrum, permits its ends to pass more freely than
the violet glass, striking effects are also obtained.[1]

This question of absorption, considered with reference to its molecular mechanism, is one of the most

---

[1] Both in foliage and in flowers we have striking differences of absorption. The copper beech and the green beech, for example, take in
different rays. But the very growth of the tree is due to some of the
rays thus taken in. Are the chemical rays, then, the same in the
copper and the green beech? In two such flowers as the primrose and
the violet, where the absorptions, to judge by the colours, are almost complementary, are the chemically active rays the same? The general relation of colour to chemical action is worthy of the application of the
method by which Dr. Draper proved so conclusively the chemical potency
of the yellow rays.

subtle and difficult in physics. We are not yet in a
condition to grapple with it, but we shall be by-and-
by. Meanwhile we may profitably glance back on
the web of relations which these experiments reveal
to us. We have in the first place in solar light an agent
of exceeding complexity, composed of innumerable
constituents, refrangible in different degrees. We find,
secondly, the atoms and molecules of bodies gifted
with the power of sifting solar light in the most various ways, and producing by this sifting the colours
observed in nature and art. To do this they must possess a molecular structure commensurate in complexity
with that of light itself. Thirdly, we have the human
eye and brain, so organized as to be able to take in and
distinguish the multitude of impressions thus generated.
The light, therefore, at starting is complex; to sift and
select it as they do natural bodies must be complex ; while
to take in the impressions thus generated, the human
eye and brain, however we may simplify our conceptions
of their action,[1] must be highly complex. Whence this
triple complexity ? If what are called material purposes were the only end to be served, a much simpler
mechanism would be sufficient. But, instead of simplicity, we have prodigality of relation and adaptation—
and this apparently for the sole purpose of enabling us
to see things robed in the splendours of colour. Would
it not seem that Nature harboured the intention of edu-
cating us for other enjoyments than those derivable
from meat and drink ? At all events, whatever Nature
meant—and it would be mere presumption to dogmatize

---

[1] Young, Helmholtz, and Maxwell reduce all differences of hue to
combinations in different proportions of three primary colours. It is
demonstrable by experiment that from the red, green, and violet *all* the
other colours of the spectrum may be obtained.

as to what she meant—we find ourselves here as
the upshot of her operations, endowed with capacities
to enjoy not only the materially useful, but endowed
with others of indefinite scope and application, which
deal alone with the beautiful and the true.

Sir Charles Wheatstone has recently drawn my attention to a work
by Christian Ernst Wünsch, Leipzig, 1792, in which the author an-
nounces the proposition that there are neither five nor seven, but only
three simple colours in white light.   Wünsch produces five spectra,
with five prisms and five small apertures, and he mixes the colours first
in pairs, and afterwards in other ways and proportions.   His result is that
' red is a *simple* colour incapable of being decomposed; that orange is compounded of intense red and weak green ; that yellow is a mixture of intense
red and intense green ; that green is a *simple* colour ; that blue is compounded of saturated green and saturated violet ; that indigo is a mixture
of saturated violet and weak green ; while violet is a pure *simple*
colour.'   He also finds that yellow and indigo blue produce *white* by
their mixture.   Yellow with bright blue (hochblau) also produce white,
which seems, however, to have a tinge of green, while the pigments of
these two colours when mixed always give a more or less beautiful
green.   Wünsch very emphatically distinguishes the mixture of pigments
from that of lights.   Speaking of the generation of yellow, he says,
I say expressly *red and green light*, because I am speaking about light-
colours (Lichtfarben), and not about pigments.'   However faulty his
theories may be, Wünsch's experiments appear in the main to be
precise and conclusive.   Nearly ten years subsequently Young adopted
red, green, and violet as the three primary colours, each of them
capable of producing three sensations, one of which, however, pre-
dominates over the two others.   Helmholtz adopts, elucidates, and
enriches this notion.   (Popular Lectures, p. 249.   The beautiful
paper of Helmholtz on the mixture of colours, translated by myself, is
published in the ' Philosophical Magazine ' for 1852.   Maxwell's excellent memoir on the Theory of Compound Colours is published in the
' Philosophical Transactions,' vol. 150, p. 57.)

LECTURE II.

WE might vary and extend our experiments on light indefinitely, and they certainly would prove us to possess a wonderful mastery over the phenomena. But the vesture of the agent only would thus be revealed, not the agent itself. The human mind, however, is so constituted and so educated as regards natural things, that it can never rest satisfied with this outward view of them. Brightness and freshness take possession of the mind when it is crossed by the light of principles, which show the facts of Nature to be organically connected.

Let us, then, inquire what this thing is that we have been generating, reflecting, refracting, and analyzing.

In doing this, we shall learn that the life of the
experimental philosopher is twofold. He lives, in his
vocation, a life of the senses, using his hands, eyes, and
ears in his experiments, but such a question as that
now before us carries him beyond the margin of the
senses. He cannot consider, much less answer, the
question, 'What is light?' without transporting himself to a world which underlies the sensible one, and
out of which, in accordance with rigid law, all optical
phenomena spring. To realize this subsensible world,
if I may use the term, the mind must possess a certain
pictorial power. It must be able to form definite
images of the things which that subsensible world contains; and to say that, if such or such a state of things
exist in that world, then the phenomena which appear
in ours must, of necessity, grow out of this state of
things. If the picture be correct, the phenomena
are accounted for; a physical theory has been enunciated which unites and explains them all.

This conception of physical theory implies, as you
perceive, the exercise of the imagination. Do not be
afraid of this word, which seems to render so many
respectable people, both in the ranks of science and
out of them, uncomfortable. That men in the ranks of
science should feel thus is, I think, a proof that they
have suffered themselves to be misled by the popular
definition of a great faculty instead of observing
its operation in their own minds. Without imagination we cannot take a step beyond the bourne of the
mere animal world, perhaps not even to the edge of
this. But, in speaking thus of imagination, I do not
mean a riotous power which deals capriciously with
facts, but a well-ordered and disciplined power, whose

sole function is to form conceptions which the intellect imperatively demands. Imagination hus exercised never really severs itself from the world of fact. This is the storehouse from which all its pictures are drawn; and the magic of its art consists, not in creating things anew, but in so changing the magnitude, position, and other relations of sensible things, as to render them fit for the requirements of the intellect in the subsensible world.[1]

Descartes imagined space to be filled with something that transmitted light *instantaneously*. Firstly, because, in his experience, no measurable interval was known to exist between the appearance of a flash of light, however distant, and its effect upon consciousness; and secondly, because as far as his experience went, no physical power is conveyed from place to place without a vehicle. But his imagination helped itself farther

---

[1] The following charming extract, bearing upon this point, was discovered and written out for me by my deeply lamented friend Dr. Bence Jones, late Hon. Secretary to the Royal Institution:

'In every kind of magnitude there is a degree or sort to which our sense is proportioned, the perception and knowledge of which is of greatest use to mankind. The same is the groundwork of philosophy; for, though all sorts and degrees are equally the object of philosophical speculation, yet it is from those which are proportioned to sense that a philosopher must set out in his inquiries, ascending or descending afterwards as his pursuits may require. He does well indeed to take his views from many points of sight, and supply the defects of sense by a well-regulated imagination; nor is he to be confined by any limit in space or time; but, as his knowledge of Nature is founded on the observation of sensible things, he must begin with these, and must often return to them to examine his progress by them. Here is his secure hold; and as he sets out from thence, so if he likewise trace not often his steps backwards with caution, he will be in hazard of losing his way in the labyrinths of Nature.'—(*Maclaurin: An Account of Sir I. Newton's Philosophical Discoveries. Written* 1728; *second edition,* 1750; pp. 18, 19.)

by illustrations drawn from the world of fact. 'When,' he says, 'one walks in darkness with staff in hand, the moment the distant end of the staff strikes an obstacle the hand feels it. This explains what might otherwise be thought strange, that the light reaches us instantaneously from the sun. I wish thee to believe that light in the bodies that we call luminous is nothing more than a very brisk and violent motion, which, by means of the air and other transparent media, is conveyed to the eye exactly as the shock through the walking stick reaches the hand of a blind man. This is instantaneous, and would be so even if the intervening distance were greater than that between earth and heaven. It is therefore no more necessary that anything material should reach the eye from the luminous object, than that something should be sent from the ground to the hand of the blind man when he is conscious of the shock of his staff.' The celebrated Robert Hooke first threw doubt upon this notion of Descartes, but afterwards substantially espoused it. The belief in instantaneous transmission was destroyed by the discovery of Rœmer referred to in our last lecture.

The case of Newton still more forcibly illustrates the position that in forming physical theories we draw for our materials upon the world of fact. Before he began to deal with light, he was intimately acquainted with the laws of elastic collision, which all of you have seen more or less perfectly illustrated on a billiard-table. As regards the collision of sensible masses, Newton knew the angle of incidence to be equal to the angle of reflection, and he also knew that experiment, as shown in our last lecture (fig. 3), had established the same law with regard to light. He thus found in his previous knowledge the

material for theoretic images. He had only to change
the magnitude of conceptions already in his mind to
arrive at the Emission Theory of Light. He supposed
light to consist of elastic particles of inconceivable
minuteness shot out with inconceivable rapidity by
luminous bodies. Such particles impinging upon smooth
surfaces were reflected in accordance with the ordinary
law of elastic collision. The fact of optical reflection
certainly occurred as if light consisted of such
particles, and this was Newton's sole justification for
introducing them.

But this is not all. In another important particular,
also, Newton's conceptions regarding the nature of
light were influenced by his previous knowledge. He
had been pondering over the phenomena of gravitation,
and had made himself at home amid the operations of
this universal power. Perhaps his mind at this time was
too freshly and too deeply imbued with these notions
to permit of his forming an unfettered judgment regarding the nature of light. Be that as it may, Newton
saw in refraction the action of an attractive force exerted on the light-particles. He carried his conception
out with the most severe consistency. Dropping vertically downwards towards the earth's surface, the motion of a body is accelerated as it approaches the earth.
Dropping in the same manner downwards on a horizontal
surface, say through air on glass or water, the velocity
of the light-particles, when they came close to the surface, was, according to Newton, also accelerated. Approaching such a surface obliquely, he supposed the
particles, when close to it, to be drawn down upon it,
as a projectile is drawn by gravity to the surface of the
earth. This deflection was, according to Newton, the

refraction seen in our last lecture (fig 4). Finally, it was supposed that differences of colour might be due to differences in the sizes of the particles. This was the physical theory of light enunciated and defended by Newton; and you will observe that it simply consists in the transference of conceptions born in the world of the senses to a subsensible world.

But, though the region of physical theory lies thus behind the world of senses, the verifications of theory occur in that world. Laying the theoretic conception at the root of matters, we determine by rigid deduction what are the phenomena which must of necessity grow out of this root. If the phenomena thus deduced agree with those of the actual world, it is a presumption in favour of the theory. If as new classes of phenomena arise they also are found to harmonize with theoretic deduction, the presumption becomes still stronger. If, finally, the theory confers prophetic vision upon the investigator, enabling him to predict the existence of phenomena which have never yet been seen, and if those predictions be found on trial to be rigidly correct, the persuasion of the truth of the theory becomes over-powering.

Thus working backwards from a limited number of phenomena, genius, by its own expansive force, reaches a conception which covers all the phenomena. There is no more wonderful performance of the intellect than this; but we can render no account of it. Like the scriptural gift of the Spirit, no man can tell whence it cometh. The passage from fact to principle is sometimes slow, sometimes rapid, and at all times a source of intellectual joy. When rapid, the pleasure is concentrated and becomes a kind of ecstasy or in-

toxication. To any one who has experienced this
pleasure, even in a moderate degree, the action of
Archimedes when he quitted the bath, and ran naked,
crying 'Eureka!' through the streets of Syracuse,
becomes intelligible.

How, then, did it fare with the Emission Theory when
the deductions from it were brought face to face with
natural phenomena? Tested by experiment, it was
found competent to explain many facts, and with transcendent ingenuity its author sought to make it account
for all. He so far succeeded, that men so celebrated
as Laplace and Malus, who lived till 1812, and Biot
and Brewster, who lived till our own time, were found
among his disciples.

Still, even at an early period of the existence of the
Emission Theory, one or two great names were found
recording a protest against it; and they furnish another
illustration of the law that, in forming theories, the
scientific imagination must draw its materials from the
world of fact and experience. It was known long ago
that sound is conveyed in waves or pulses through the
air; and no sooner was this truth well housed in the
mind than it was transformed into a theoretic conception. It was supposed that light, like sound, might
also be the product of wave-motion. But what, in this
case, could be the material forming the waves? For the
waves of sound we have the air of our atmosphere; but
the stretch of imagination which filled all space with a
luminiferous ether trembling with the waves of light
was so bold as to shock cautious minds. In one of my
latest conversations with Sir David Brewster he said to
me that his chief objection to the undulatory theory
of light was that he could not think the Creator guilty

of so clumsy a contrivance as the filling of space with ether in order to produce light. This, I may say, is very dangerous ground, and the quarrel of science with Sir David, on this point, as with many estimable persons on other points, is, that they profess to know too much about the mind of the Creator.

This conception of an ether was advocated, and indeed applied to various phenomena of optics, by the celebrated astronomer, Huyghens. It was espoused and defended by the celebrated mathematician, Euler. They were, however, opposed by Newton, whose authority at the time bore them down. Or shall we say it was authority merely? Not quite so. Newton's preponderance was in some degree due to the fact that, though Huyghens and Euler were right in the main, they did not possess sufficient data to *prove* themselves right. No human authority, however high, can maintain itself against the voice of Nature speaking through experiment. But the voice of Nature may be an uncertain voice, through the scantiness of data. This was the case at the period now referred to, and at such a period by the authority of Newton all antagonists were naturally overborne.

Still, this great Emission Theory, which held its ground so long, resembled one of those circles which, according to your countryman Emerson, the force of genius periodically draws round the operations of the intellect, but which are eventually broken through by pressure from behind. In the year 1773 was born, at Milverton, in Somersetshire, one of the most remarkable men that England ever produced. He was educated for the profession of a physician, but was too strong to be tied down to professional routine. He devoted himself to the study of natural philosophy, and became in all its departments a master. He was also a master of letters. Languages, ancient and modern, were housed within his brain and, to use the words of his epitaph, ' he first penetrated the obscurity which had veiled for ages the hieroglyphics of Egypt.' It fell to the lot of this man to discover facts in optics which Newton's theory was incompetent to explain, and his mind roamed in search of a sufficient theory. He had made himself acquainted with all the phenomena of wave-motion; with all the phenomena of sound; working successfully in this domain as an original discoverer. Thus informed and disciplined, he was prepared to detect any resemblance which might reveal itself between the phenomena of light and those of wave-motion. Such resemblances he did detect; and, spurred on by the discovery, he pursued his speculations and his experiments, until he finally succeeded in placing on an immovable basis the Undulatory Theory of Light.

The founder of this great theory was Thomas Young, a name, perhaps, unfamiliar to many of you, but which ought to be familiar to you all. Permit me, therefore, by a kind of geometrical construction which I once ventured to employ in London, to give you a notion of the magnitude of this man. Let Newton stand erect in his age, and Young in his. Draw a straight line from Newton to Young, tangent to the heads of both. This line would slope downwards from Newton to Young, because Newton was certainly the taller man of the two. But the slope would not be steep, for the difference of stature was not excessive. The line would form what engineers call a gentle gradient from Newton to Young. Place underneath this line the biggest man born in the interval between both. It may be doubted whether he would reach the line; for if he did he would be taller intellectually than Young, and there was probably none taller. But I do not want you to rest on English estimates of Young; the German, Helmholtz, a kindred genius, thus speaks of him : 'His was one of the most profound minds that the world has ever seen ; but he had the misfortune to be too much in advance of his age. He excited the wonder of his contemporaries, who, however, were unable to follow him to the heights at which his daring intellect was accustomed to soar. His most important ideas lay, therefore, buried and forgotten in the folios of the Royal Society, until a new generation gradually and painfully made the same discoveries, and proved the exactness of his assertions and the truth of his demonstrations.'

It is quite true, as Helmholtz says, that Young was in advance of his age; but something is to be added which illustrates the responsibility of our public writers. For twenty years this man of genius was quenched—hidden from the appreciative intellect of his countrymen—deemed in fact a dreamer, through the vigorous sarcasm of a writer who had then possession of the public ear, and who in the *Edinburgh Review* poured ridicule upon Young and his speculations. To the celebrated Frenchmen Fresnel and Arago, he was first indebted for the restitution of his rights, for they, especially Fresnel, remade independently, as Helmholtz says, and vastly extended his discoveries. To the students of his works Young has long since appeared in his true light, but these twenty blank years pushed him from the public mind, which became in turn

filled with the fame of Young's colleague at the Royal
Institution, Davy, and afterwards with the fame of
Faraday. Carlyle refers to a remark of Novalis, that
a man's self-trust is enormously increased the moment
he finds that others believe in him. If the opposite
remark be true—if it be a fact that public disbelief
weakens a man's force—there is no calculating the
amount of damage these twenty years of neglect may
have done to Young's productiveness as an investigator. It remains to be stated that his assailant was
Mr. Henry Brougham, afterwards Lord Chancellor of
England.

Our hardest work is now before us. But the
capacity for hard work depends in a great measure
on the antecedent winding up of the will; I would
call upon you, therefore, to gird up your loins for our
coming labours. If we succeed in climbing the hill
which faces us to-night, our future difficulties will not
be insurmountable.

In the earliest writings of the ancients we find the
notion that sound is conveyed by the air. Aristotle
gives expression to this notion, and the great architect
Vitruvius compares the waves of sound to waves of
water. But the real mechanism of wave-motion was
hidden from the ancients, and indeed was not made
clear until the time of Newton. The central difficulty
of the subject was, to distinguish between the motion
of the wave itself and the motion of the particles which
at any moment constitute the wave.

Stand upon the sea-shore and observe the advancing
rollers before they are distorted by the friction of the
bottom. Every wave has a back and a front, and, if

you clearly seize the image of the moving wave, you will see that every particle of water along the front of the wave is in the act of rising, while every particle along its back is in the act of sinking. The particles in front reach in succession the crest of the wave, and as soon as the crest is passed they begin to fall. They then reach the furrow or *sinus* of the wave, and can sink no farther. Immediately afterwards they become the front of the succeeding wave, rise again until they reach the crest, and then sink as before. Thus, while the waves pass onward horizontally, the individual particles are simply lifted up and down vertically. Observe a sea-fowl, or, if you are a swimmer, abandon yourself to the action of the waves; you are not carried forward, but simply rocked up and down. The propagation of a wave is the propagation of *a form*, and not the transference of the substance which constitutes the wave.

The *length* of the wave is the distance from crest to crest, while the distance through which the individual particles oscillate is called the *amplitude* of the oscillation. You will notice that in this description the particles of water are made to vibrate *across* the line of propagation.[1]

And now we have to take a step forwards, and it is the most important step of all. You can picture two series of waves proceeding from different origins through the same water. When, for example, you

---

[1] I do not wish to encumber the conception here with the details of the motion, but I may draw attention to the beautiful model of Prof. Lyman, wherein waves are shown to be produced by the *circular* motion of the particles. This, as proved by the brothers Weber, is the real motion in the case of water-waves.

throw two stones into still water, the ring-waves proceeding from the two centres of disturbance intersect
each other. Now, no matter how numerous these waves
may be, the law holds good that the motion of every
particle of the water is the algebraic sum of all the
motions imparted to it. If crest coincide with crest
and furrow with furrow, the wave is lifted to a double
height above its sinus; if furrow coincide with crest, the
motions are in opposition, and their sum is zero.
We have then *still* water, which we shall learn presently corresponds to what we call *darkness* in reference
to our present subject. This action of wave upon
wave is technically called *interference*, a term to be
remembered.

To the eye of a person conversant with these principles, nothing can be more interesting than the crossing
of water ripples. Through the interference of the
waves, the intersecting surface is sometimes shivered
into the most beautiful mosaic, trembling rhythmically
as if with a kind of visible music. When waves are
skilfully generated in a dish of mercury, a strong
light thrown upon the shining surface, and reflected
on to a screen, reveals the motions of the liquid
metal. The shape of the vessel determines the forms
of the figures produced. In a circular dish, for example, a disturbance at the centre propagates itself as
a series of circular waves, which, after reflection, again
meet at the centre. If the point of disturbance be a
little way removed from the centre, the interference of
the direct and reflected waves produces the magnificent
chasing shown in the annexed figure.[1] The light

---

[1] Copied from Weber's *Wellenlehre.*

reflected from such a surface yields a pattern of extraordinary beauty.   When the mercury is slightly struck

Fig. 10.

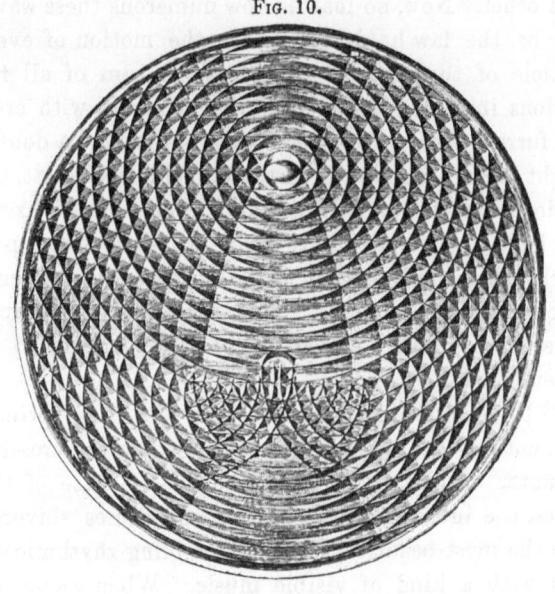

by a needle-point in a direction concentric with the
surface of the vessel, the lines of light run round in
mazy coils, interlacing and unravelling themselves
in a wonderful manner.   When the vessel is square,
a splendid checker-work is produced by the crossing of
the direct and reflected waves.   Thus, in the case of
wave-motion, the most ordinary causes give rise to the
most exquisite effects.   The words of your countryman,
Emerson, are perfectly applicable here :—

> ' Thou can'st not wave thy staff in the air,
>   Or dip thy paddle in the lake,
> But it carves the brow of beauty there,
>   And the ripples in rhymes the oars forsake.'

The most impressive illustration of the action of waves on waves that I have ever seen occurs near Niagara. For a distance of two miles, or thereabouts, below the Falls, the river Niagara flows unruffled through its excavated gorge. The bed subsequently narrows and deepens, and the water consequently quickens its motion. At the place called the 'Whirlpool Rapids,' I estimated the width of the river at 300 feet, an estimate confirmed by the dwellers on the spot. When it is remembered that the drainage of nearly half a continent is compressed into this space, the impetuosity of the river's escape through this gorge may be imagined.

Two kinds of motion are here obviously active, a motion of translation and a motion of undulation—the race of the river through its gorge, and the great waves generated by its collision with the obstacles in its way. In the middle of the stream, the rush and tossing are most violent; at all events, the impetuous force of the individual waves is here most strikingly displayed. Vast pyramidal heaps leap incessantly from the river, some of them with such energy as to jerk their summits into the air, where they hang suspended as bundles of liquid pearls, which, when shone upon by the sun, are of undescribable beauty.

The first impression, and, indeed, the current explanation of these Rapids is, that the central bed of the river is cumbered with large boulders, and that the jostling, tossing, and wild leaping of the water there are due to its impact against these obstacles. A very different explanation occurred to me upon the spot. Boulders derived from the adjacent cliffs visibly cumber the *sides* of the river. Against these the water rises and sinks rhythmically but violently, large waves being thus produced. On the generation of each wave there is an immediate compounding of the wave motion with the river motion. The ridges, which in still water would proceed in circular curves round the centre of disturbance, cross the river obliquely, and the result is that at the centre waves commingle which have really been generated at the sides. This crossing of waves may be seen on a small scale in any gutter after rain; it may also be seen on simply pouring water from a wide-lipped jug. In the first instance, then, we had a composition of wave motion with river motion; here we have the coalescence of waves with waves. Where crest and furrow cross each other, the motion is annulled; where furrow and furrow cross, the river is ploughed to a greater depth; and where crest and crest aid each other, we have that astonishing leap of the water which breaks the cohesion of the crests, and tosses them shattered into the air. The phenomena observed at the Whirlpool Rapids constitute in fact, one of the grandest illustrations of the principle of interference.

Thomas Young's fundamental discovery in optics was that the principle of Interference applied to light. Long prior to his time an Italian philosopher, Grimaldi, had stated that under certain circumstances two thin beams of light, each of which, acting singly, produced a luminous spot upon a white wall, when caused to act together, partially quenched each other and darkened the spot. This was a statement of fundamental significance, but it required the discoveries and the genius of Young to give it meaning. How he did so will gradually become clear to you. You know that air is

compressible; that by pressure it can be rendered more
dense, and that by dilatation it can be rendered more
rare. Properly agitated, a tuning-fork now sounds in
a manner audible to you all, and most of you know that
the air through which the sound is passing is parcelled
out into spaces in which the air is condensed, followed
by other spaces in which the air is rarefied. These
condensations and rarefactions constitute what we call
waves of sound. You can imagine the air of a room
traversed by a series of such waves, and you can imagine
a second series sent through the same air, and so related
to the first that condensation coincides with condensation and rarefaction with rarefaction. The consequence
of this coincidence would be a louder sound than that
produced by either system of waves taken singly. But
you can also imagine a state of things where the condensations of the one system fall upon the rarefactions
of the other system. In this case the two systems
would completely neutralize each other. Each of them
taken singly produces sound; both of them taken
together produce no sound. Thus, by adding sound
to sound we produce silence, as Grimaldi in his experiment produced darkness by adding light to light.

The analogy between sound and light here flashes upon
the mind. Young generalized this observation. He
discovered a multitude of similar cases, and determined
their precise conditions. On the assumption that
light was wave-motion, all his experiments on interference were explained; on the assumption that light
was flying particles, nothing was explained. In the
time of Huyghens and Euler a medium had been
assumed for the transmission of the waves of light;
but Newton raised the objection that, if light consisted

of the waves of such a medium, shadows could not exist. The waves, he contended, would bend round opaque bodies and produce the motion of light behind them, as sound turns a corner, or as waves of water wash round a rock. It was proved that the bending round referred to by Newton actually occurs, but that the inflected waves abolish each other by their mutual interference. Young also discerned a fundamental

Fig. 11.

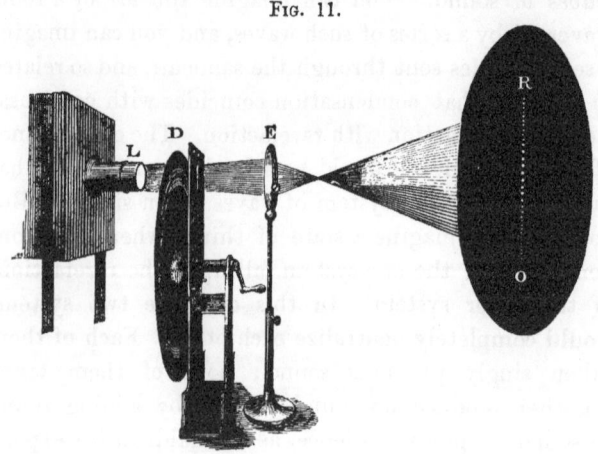

difference between the waves of light and those of sound. Could you see the air through which sound-waves are passing, you would observe every individual particle of air oscillating to and fro in the direction of propagation. Could you see the luminiferous ether, you would also find every individual particle making a small excursion to and fro, but here the motion, like that assigned to the water-particles above referred to, would be *across* the line of propagation. The vibrations of the air are *longitudinal*, the vibrations of the ether are *transversal*.

It is my desire that you should realize with clearness
the character of wave-motion, both in ether and in air.
And, with this view, I bring before you an experiment
wherein the air-particles are represented by small spots
of light (R O, fig. 11). They are derived from a clean
spiral, drawn upon a circle of blackened glass (D), so
that when the circle rotates, the spots move in successive
pulses over the screen.[1] In this experiment you have

Fig. 12.

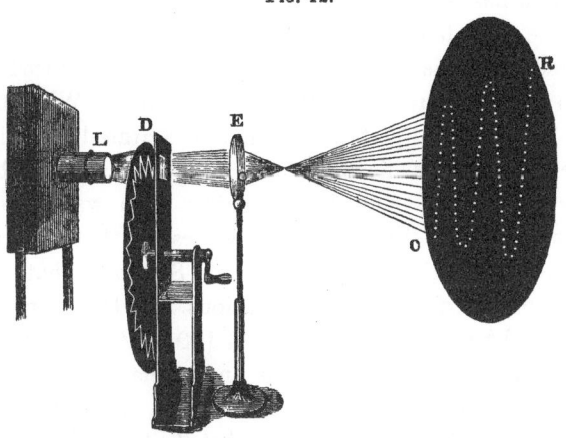

clearly set before you how the pulses travel incessantly
forward, while their component particles perform oscillations to and fro. This is the picture of a sound-wave, in
which the vibrations are longitudinal. By another glass
wheel (D, fig. 12) we produce an image of a transverse
wave (O R), and here we observe the waves travelling in
succession over the screen, while each individual spot of
light performs an excursion to and fro across the line of
propagation.

[1] The apparatus is constructed by that excellent acoustic mechanician,
M. Rudolf König, of Paris.

Notice what follows when the glass wheel is turned very quickly. Objectively considered, the transverse waves propagate themselves as before, but subjectively the effect is totally changed. Because of the retention of impressions upon the retina, the spots of light simply describe a series of parallel luminous lines upon the screen, the length of these lines marking the amplitude of the vibration. Here the impression of wave-motion has totally disappeared.

The most familiar illustration of the interference of sound-waves is furnished by the *beats* produced by two musical sounds slightly out of unison. When two tuning-forks in perfect unison are agitated together the two sounds flow without roughness, as if they were but one. But, by attaching with wax to one of the forks a little weight, we cause it to vibrate more slowly than its neighbour. Suppose that one of them performs 101 vibrations in the time required by the other to perform 100, and suppose that at starting the condensations and rarefactions of both forks coincide. At the 101st vibration of the quickest fork they will again coincide, that fork at this point having gained one whole vibration, or one whole wavelength upon the other. But a little reflection will make it clear that, at the 50th vibration, the two forks are in opposition; here the one tends to produce a condensation where the other tends to produce a rarefaction; by the united action of the two forks, therefore, the sound is quenched, and we have a pause of silence. This occurs where one fork has gained *half a wavelength* upon the other. At the 101st vibration, as already stated, we have coincidence, and, therefore, augmented sound; at the 150th vibration we have

again a quenching of the sound. Here the one fork is
three half-waves in advance of the other. In general
terms, the waves conspire when the one series is an
even number of half-wave lengths, and they are destroyed when the one series is an *odd* number of half-
wave lengths in advance of the other. With two forks
so circumstanced, we obtain those intermittent shocks
of sound separated by pauses of silence, to which we
give the name of beats. By a suitable arrangement,
moreover, it is possible to make one sound wholly
extinguish another. Along four distinct lines, for
example, the vibrations of the two prongs of a tuning-
fork completely blot each other out.[1]

The *pitch* of sound is wholly determined by the
rapidity of the vibration, as the *intensity* is by the amplitude. The rise of pitch with the rapidity of the
impulses may be illustrated by the syren, which consists of a perforated disk rotating over a cylinder into
which air is forced, and the end of which is also perforated. When the perforations of the disk coincide
with those of the cylinder, a puff escapes ; and, when
the puffs succeed each other with sufficient rapidity,
the impressions upon the auditory nerve link themselves together to a continuous musical note. The more
rapid the rotation of the disk the quicker is the succession of the impulses, and the higher the pitch of the
note. Indeed, by means of the syren the number of
vibrations due to any musical note, whether it be that of
an instrument or of the human voice, may be accurately
determined.

In front of our lamp now stands a homely instrument (S, fig. 13) of this character. The perforated

---

[1] Sound : Lecture VII. Longmans.

disk is turned by a wheel and band, and, when the
two sets of perforations coincide, a series of spots of

FIG. 13.

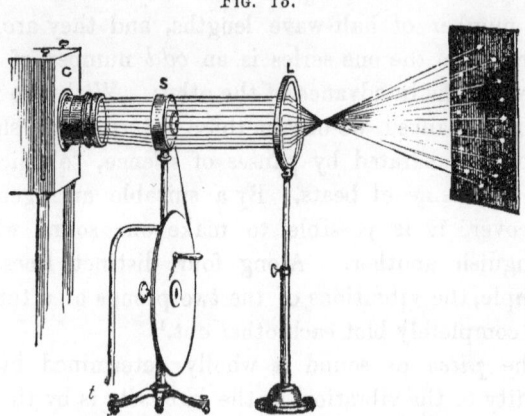

light, sharply defined by the lens L, ranged on the
circumference of a circle, is seen upon the screen.   On
slowly turning the disk, a flicker is produced by the
alternate stoppage and transmission of the light.   At
the same time air is urged into the syren through the
tube *t*, and you hear a fluttering sound corresponding
to the flickering light.   But, by augmenting the
rapidity of rotation, the light, though intercepted as
before, appears perfectly steady, through the persistence of impressions upon the retina ; and, about the
time when the optical impression becomes continuous,
the auditory impression becomes equally so ; the puffs
from the syren linking themselves then together to a
continuous musical note, which rises in pitch with the
rapidity of the rotation.   A movement of the head
causes the image of the spots to sweep over the retina,
producing beaded lines : the same effect is produced

upon our screen by the sweep of a looking-glass which has received the thin beams from the syren.

What pitch is to the ear in acoustics, colour is to the eye in the undulatory theory of light. Though never seen, the lengths of the waves of light have been determined. Their existence is proved *by their effects*, and from their effects also their lengths may be accurately deduced. This may, moreover, be done in many ways, and, when the different determinations are compared, the strictest harmony is found to exist between them. This consensus of evidence is one of the strongest points of the undulatory theory. The shortest waves of the visible spectrum are those of the extreme violet; the longest, those of the extreme red; while the other colours are of intermediate pitch or wave-length. The length of a wave of the extreme red is such that it would require 36,918 of them placed end to end to cover one inch, while 64,631 of the extreme violet waves would be required to span the same distance.

Now, the velocity of light, in round numbers, is 190,000 miles per second. Reducing this to inches, and multiplying the number thus found by 36,918, we obtain the number of waves of the extreme red in 190,000 miles. *All these waves enter the eye, and strike the retina at the back of the eye in one second.* The number of shocks per second necessary to the production of the impression of red is, therefore, four hundred and fifty-one millions of millions. In a similar manner, it may be found that the number of shocks corresponding to the impression of violet is seven hundred and eighty-nine millions of millions.

All space is filled with matter oscillating at such rates. From every star waves of these dimensions

move with the velocity of light like spherical shells
outwards.   And in the ether, just as in the water, the
motion of every particle is the algebraic sum of all
the separate motions imparted to it.   Still, one motion does not blot the other out; or, if extinction
occur at one point, it is strictly atoned for at some other
point.   Every star declares by its light its undamaged
individuality, as if it alone had sent its thrills through
space.

The principle of interference, as proved by Young,
applies to the waves of light as it does to the waves
of water and the waves of sound.   And the conditions
of interference are the same in all three.   If two
series of light-waves of the same length start at the
same moment from a common origin (say A, fig.14),

Fig. 14.

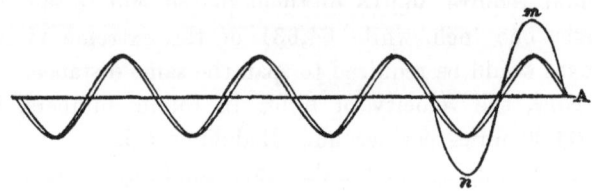

crest coincides with crest, sinus with sinus, and the two
systems blend together to a single system (A *m n*) of
double amplitude.   If both series start at the same
moment, one of them being, at starting, a whole wave-
length in advance of the other, they also add themselves together, and we have an augmented luminous
effect.   Just as in the case of sound, the same occurs
when the one system of waves is any *even* number of
semi-undulations in advance of the other.   But if the
one system be half a wave-length (as at A' *a'* fig.15), or
any *odd* number of half wave-lengths in advance, then

the crests of the one fall upon the sinuses of the other; the one system, in fact, tends to *lift* the particles of

Fig. 15.

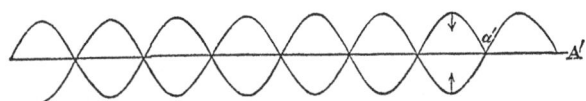

ether at the precise places where the other tends to *depress* them; hence, through the joint action of these opposing forces (indicated by the arrows) the light-ether remains perfectly still. This stillness of the ether is what we call darkness, which corresponds, as already stated, with a dead level in the case of water.

It was said in our first lecture, with reference to the colours produced by absorption, that the function of natural bodies is selective, not creative; that they extinguish certain constituents of the white solar light, and appear in the colours of the unextinguished light. It must at once flash upon your minds that, inasmuch as we have in interference an agency by which light may be self-extinguished, we may have in it the conditions for the production of colour. But this would imply that certain constituents are quenched by interference, while others are permitted to remain. This is the fact; and it is entirely due to the difference in the lengths of the waves of light.

This subject may be illustrated by the class of phenomena which first suggested the undulatory theory to the mind of Hooke. These are the colours of thin transparent films of all kinds, known as the *colours of thin plates*. In this relation no object in the world possesses a deeper scientific interest than a common soap-bubble. And here let me say emerges one of the difficulties which the student of pure science encounters in the presence of 'practical' communities like those of America and England; it is not to be expected that such communities can entertain any profound sympathy with labours which seem so far removed from the domain of practice as many of the labours of the man of science are. Imagine Dr. Draper spending his days in blowing soap-bubbles and in studying their colours! Would you show him the necessary patience, or grant him the necessary support? And yet be it remembered it was thus that minds like those of Boyle, Newton and Hooke were occupied; and that on such experiments has been founded a theory, the issues of which are incalculable. I see no other way for you, laymen, than to trust the scientific man with the choice of his inquiries; he stands before the tribunal of his peers, and by their verdict on his labours you ought to abide.

Whence, then, are derived the colours of the soap-bubble? Imagine a beam of white light impinging on the bubble. When it reaches the first surface of the film, a known fraction of the light is reflected back. But a large portion of the beam enters the film, reaches its second surface, and is again in part reflected. The waves from the second surface thus turn back and hotly pursue the waves from the first surface. And, if the thickness of the film be such as to cause the necessary retardation, the two systems of waves interfere with each other, producing augmented or diminished light, as the case may be.

But, inasmuch as the waves of light are of different lengths, it is plain that, to produce self-extinction in the case of the longer waves, a greater thickness of film is necessary than in the case of the shorter ones.

Different colours, therefore, must appear at different thicknesses of the film.

Take with you a little bottle of spirit of turpentine, and pour it into one of your country ponds. You will then see the flashing of those colours over the surface of the water. On a small scale we produce them thus: A common tea-tray is filled with water, beneath the surface of which dips the end of a pipette. A beam of light falls upon the water, and is reflected by it to the screen. Spirit of turpentine is poured into the pipette; it descends, issues from the end in minute drops, which rise in succession to the surface. On reaching it, each drop spreads suddenly out as a film, and glowing colours immediately flash forth upon the screen. The colours change as the thickness of the film changes by evaporation. They are also arranged in zones in consequence of the gradual diminution of thickness from the centre outwards.

Any film whatever will produce these colours. The film of air between two plates of glass squeezed together, exhibits, as shown by Hooke, rich fringes of colour. A particularly fine example of these fringes is now before you. Nor is even air necessary; the rupture of optical continuity suffices. Smite with an axe the black, transparent ice—black, because it is pure and of great depth —under the moraine of a glacier; you readily produce in the interior flaws which no air can reach, and from these flaws the colours of thin plates sometimes break like fire. But the origin of most historic interest is, as already stated, the soap-bubble. With one of these mixtures employed by the eminent blind philosopher Plateau in his researches on the cohesion figures of thin films, we obtain in still air a bubble ten or twelve inches in diameter. You may look at the bubble itself, or you may look at its projection upon the screen, rich colours arranged in zones are, in both cases, exhibited. Rendering the beam parallel, and permitting it to impinge upon the sides, bottom, and top, of the bubble, gorgeous fans of colour overspread the screen, which rotate as the beam is carried round the circumference of the bubble. By this experiment the internal motions of the film are also strikingly displayed.

Not in a moment are great theories elaborated : the facts which demand them are first called into prominence by observant minds ; then to the period of observation succeeds a period of pondering and of tentative explanation. By such efforts the human mind is gradually prepared for the final theoretic illumination. The colours of thin plates, for example, occupied the attention of the celebrated Robert Boyle. In his 'Experimental History of Colours' he contends against the schools which affirmed that colour was 'a penetrative quality that reaches to the innermost parts of the object,' adducing opposing facts. 'To give you a first instance,' he says, 'I shall need but to remind you of what I told you a little after the beginning of this essay, touching the blue and red and yellow that may be produced upon a piece of tempered steel; for these colours, though they be very vivid, yet if you break the steel they adorn they will appear to be but superficial.' He then describes, in phraseology which shows the delight he took in his work, the following beautiful experiment :—

'We took a quantity of clean lead, and melted it with a strong fire, and then immediately pouring it out into a clean vessel of convenient shape and matter

(we used one of iron, that the great and sudden heat
might not injure it), and then carefully and nimbly
taking off the scum that floated on the top, we perceived, as we expected, the smooth and glossy surface
of the melted matter to be adorned with a very glorious
colour, which being as transitory as delightful, did
almost immediately give place to another vivid colour,
and that was as quickly succeeded by a third, and this,
as it were, chased away by a fourth, and so these wonderfully vivid colours successively appeared and vanished
till the metal ceasing to be hot enough to hold any
longer this pleasing spectacle, the colours that chanced
to adorn the surface when the lead thus began to cool
remained upon it, but were so superficial that how
little soever we scraped off the surface of the lead, we
did, in such places, scrape off all the colour.' ' These
things,' he adds, ' suggested to me some thoughts or
ravings which I have not now time to acquaint you
with.' [1]

He extends his observations to chemical essential
oils and spirit of wine, ' which being shaken till
they have good store of bubbles, those bubbles will (if
attentively considered) appear adorned with various and
lovely colours, which all immediately vanish upon the
retrogressing of the liquid which affords these bubbles
their skins into the rest of the oil.' He also refers to the
colours of the soap bubble and to those of glass films.
' I have seen one that was skilled in fashioning glasses
by the help of a lamp blowing some of them so strongly
as to burst them ; whereupon it was found that the
tenacity of the metal was such that before it broke it

---

[1] Boyle's Works, Birch's edition, p. 675.

suffered itself to be reduced into films so extremely
thin that they constantly showed upon their surfaces
the varying colours of the rainbow.'[1]

Subsequent to Boyle the colours of thin plates
occupied the attention of the celebrated Robert
Hooke, in whose writings we find a. dawning of the
undulatory theory. He describes with great distinctness the colours obtained with thin flakes of 'Muscovy
glass' (talc), also those surrounding flaws in crystals
where optical continuity is destroyed. He shows very
clearly the dependence of the colour upon the thickness of the film, and proves by microscopic observation
that plates of a uniform thickness yield uniform colours.
' If,' he says, 'you take any small piece of the Muscovy
glass, and with a needle, or some other convenient
instrument, cleave it oftentimes into thinner and thinner laminæ, you shall find that until you come to a
determinate thinness of them they shall appear transparent and colourless, but if you continue to split and
divide them further, you shall find at last that each
plate shall appear most lovely tinged or imbued with
a determinate colour. If, further, by any means you
so flaw a pretty thick piece that one part begins to
cleave a little from the other, and between these two
there be gotten some pellucid medium, those laminated
or pellucid bodies that fill that space shall exhibit
several rainbows or coloured lines, the colours of which
will be disposed and ranged according to the various
thicknesses of the several parts of the plate.' He then
describes fully and clearly the experiment with pressed
glasses already referred to :—

[1] Page 743.

'Take two small pieces of ground and polished look-
ing-glass-plate, each about the bigness of a shilling :
take these two dry, and with your forefingers and
thumbs press them very hard and close together, and
you shall find that when they approach each other
very near there will appear several irises or coloured
lines, in the same manner almost as in the Muscovy
glass ; and you may very easily change any of the
colours of any part of the interposed body by pressing
the plates closer and harder together, or leaving them
more lax—that is, a part which appeared coloured with
a red, may be presently tinged with a yellow, blew,
green, purple, or the like.   Any substance,' he says,
' provided it be thin and transparent, will show these
colours.'   Like Boyle, he obtained them with glass
films ; he also ' produced them with bubbles of pitch,
rosin, colophony, turpentine, solutions of several gums,
as gum arabic in water, any glutinous liquor, as wort,
wine, spirit of wine, oyl of turpentine, glare of snails,
&c.'

Hooke's writings show that even in his day the idea that
both light and heat are modes of motion had taken possession of many minds.   ' First,' he says, ' that all kind of
fiery burning bodies have their parts in motion I think
will be very easily granted me.   That the spark struck
from a flint and steel is in rapid agitation I have elsewhere made probable. . . . . that heat argues a motion
of the internal parts is (as I said before) generally
granted. . . . . . and that in all extremely hot shining
bodies there is a very quick motion that causes light,
as well as a more robust that causes heat, may be
argued from the celerity wherewith the bodies are dissolved.   Next, it must be *a vibrative motion.*'   His

reference to the quick motion of light and the more robust motion of heat is a remarkable stroke of sagacity; but Hooke's direct insight is better than his reasoning; for the proofs he adduces that light is 'a vibrating motion' have no particular bearing upon the question.

Still the Undulatory Theory was undoubtedly dawning upon the mind of this remarkable man. In endeavouring to account for the colours of thin plates, he again refers to the relation of colour to thickness: he dwells upon the fact that the film which shows these colours must be transparent, proving this by showing that however thin an opaque body was rendered no colours were produced. 'This,' he says, 'I have often tried by pressing a small globule of mercury between two smooth plates of glass, whereby I have reduced that body to a much greater thinness than was requisite to exhibit the colours with a transparent body.' Then follows the sagacious remark that to produce the colours 'there must be a considerable reflecting body adjacent to the under or further side of the lamina or plate: for this I always found, that the greater that reflection was, the more vivid were the appearing colours. From which observations,' he continues, 'it is most evident, *that the reflection from the under or further side of the body is the principal cause of the production of these colours.*'

He draws a diagram, correctly representing the reflection at the two surfaces of the film, but here his clearness ends. He ascribes the colours to a coalescence or confusion of the two reflected pulses; the principle of interference being unknown to him he could not go further in the way of explanation. He adds, however, a remark of the utmost importance.

'One thing which seems of the greatest concern in this hypothesis is to determine the greatest and least thickness requisite for these effects, which, though I have not been wanting in attempting, yet so exceeding thin are these coloured plates, and so imperfect our microscope, that I have not been hitherto successful.'

In this way, then, by the active operation of different minds, facts are observed, examined, and the precise conditions of their appearance determined. All such work in science is the prelude to other work ; and the efforts of Boyle and Hooke cleared the way for the optical career of Newton. He conquered the difficulty which Hooke had found insuperable, and determined by accurate measurements the relation of the thickness of the film to the colour of displays. In doing this his first care was to obtain a film of variable and calculable

Fig. 16.

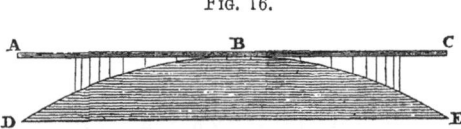

depth. On a plano-convex glass lens (D B E, fig. 16) of very feeble curvature he laid a plate of glass (A C) with a plane surface, thus obtaining a film of air of gradually increasing depth from the point of contact (B) outwards. On looking at the film in monochromatic light he saw, with the delight attendant on fulfilled prevision, surrounding the place of contact a series of bright rings separated from each other by dark ones, and becoming more closely packed together as the distance from the point of contact augmented

(as in fig. 17). When he employed red light, his rings
had certain diameters; when he employed blue light,

FIG. 17.

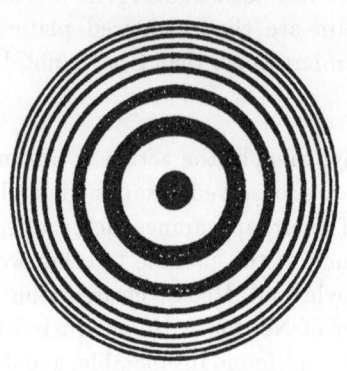

the diameters were less. In general terms, the more
refrangible the light the smaller were the rings.
Causing his glasses to pass through the spectrum
from red to blue, the rings gradually contracted;
when the passage was from blue to red, the
rings expanded. This is a beautiful experiment, and
appears to have given Newton the most lively satisfaction. When white light fell upon the glasses,
inasmuch as the colours were not superposed, a series
of *iris-coloured* circles were obtained. A magnified
image of *Newton's rings* is now before you, and, by
employing in succession red, blue, and white light, we
obtain all the effects observed by Newton. You notice
that in monochromatic light the rings run closer and
closer together as they recede from the centre. This is
due to the fact that at a distance the film of air thickens
more rapidly than near the centre. When white light

is employed, this closing up of the rings causes the various colours to be superposed, so that after a certain thickness they are blended together to white light, the rings then ceasing altogether. It needs but a moment's reflection to understand that the colours of thin plates are never unmixed and monochromatic.

Newton compared the tints obtained in this way with the tints of his soap-bubble, and he calculated the corresponding thickness. How he did this may be thus made plain to you: Suppose the water of the ocean to be absolutely smooth; it would then accurately represent the earth's curved surface. Let a perfectly horizontal plane touch the surface at any point. Knowing the earth's diameter, any engineer or mathematician in this room could tell you how far the sea's surface will lie below this plane, at the distance of a yard, ten yards, a hundred yards, or a thousand yards from the point of contact of the plane and the sea. It is common, indeed, in levelling operations, to allow for the curvature of the earth. Newton's calculation was precisely similar. His plane glass was a tangent to his curved one. From its refractive index and focal distance he determined the diameter of the sphere of which his curved glass formed a segment, he measured the distances of his rings from the place of contact, and he calculated the depth between the tangent plane and the curved surface, exactly as the engineer would calculate the distance between his tangent plane and the surface of the sea. The wonder is, that, where such infinitesimal distances are involved, Newton, with the means at his disposal, could have worked with such marvellous exactitude.

To account for these rings was the greatest difficulty

that Newton ever encountered.   He quite appreciated
the difficulty. Over his eagle-eye there was no film—no
vagueness in his conceptions.   At the very outset his
theory was confronted by the question, Why, when a
beam of light is incident on a transparent body, are
some of the light-particles reflected and some transmitted?   Is it that there are two kinds of particles,
the one specially fitted for transmission and the other
for reflection?   This cannot be the reason; for, if
we allow a beam of light which has been reflected
from one piece of glass to fall upon another, it, as a
general rule, is also divided into a reflected and a transmitted portion.   Thus the particles once reflected are
not always reflected, nor are the particles once transmitted always transmitted.   Newton saw all this; he
knew he had to explain why it is that the self-same
particle is at one moment reflected and at the next
moment transmitted.   It could only be through *some
change in the condition of the particle itself*.   The
self-same particle, he affirmed, was affected by 'fits'
of easy transmission and reflection.

If you are willing to follow me in an attempt to
reveal the speculative ground-work of this theory of
fits, the intellectual discipline will, I think, repay you
for the necessary effort of attention.   Newton was chary
of stating what he considered to be the cause of the
fits, but there can hardly be a doubt that his mind
rested on a physical cause.   Nor can there be a doubt
that here, as in all attempts at theorising, he was
compelled to fall back upon experience for the materials
of his theory.   Let us attempt to restore his course of
thought and observation.   A magnet would furnish

him with the notion of attracted and repelled poles;
and he who habitually saw in the visible an image of
the invisible would naturally endow his light-particles
with such poles. Turning their attracted poles towards
a transparent substance, the particles would be sucked
in and transmitted; turning their repelled poles, they
would be driven away or reflected. Thus, by the
ascription of poles, the transmission and reflection of
the self-same particle at different times might be
accounted for.

Regard these rings of Newton as seen in pure red
light: they are alternately bright and dark. The
film of air corresponding to the outermost of them
is not thicker than an ordinary soap-bubble, and it
becomes thinner on approaching the centre; still
Newton, as I have said, measured the thickness corresponding to every ring, and showed the difference
of thickness between ring and ring. Now, mark the
result. For the sake of convenience, let us call the
thickness of the film of air corresponding to the first
dark ring d; then Newton found the distance corresponding to the second dark ring $2\,d$; the thickness
corresponding to the third dark ring $3\,d$; the thickness corresponding to the tenth dark ring $10\,d$, and so
on. Surely there must be some hidden meaning in this
little distance d, which turns up so constantly? One
can imagine the intense interest with which Newton
pondered its meaning. Observe the probable outcome
of his thought. He had endowed his light-particles
with poles, but now he is forced to introduce the notion
of *periodic recurrence*. Here his power of transfer
from the sensible to the subsensible would render it
easy for him to suppose the light-particles animated,

not only with a motion of translation, but also with ·a
motion of rotation.   Newton's astronomical knowledge
rendered all such conceptions familiar to him.   The
earth has such a double motion.   In the time occupied
in passing over a million and a half of miles of its
orbit—that is, in twenty-four hours—our planet performs a complete rotation, and, in the time required to
pass over the distance d, Newton's light-particle must
be supposed to perform a complete rotation.   True, the
light-particle is smaller than the planet, and the distance d, instead of being a million and a half of miles,
is a little over the ninety thousandth of an inch.   But
the two conceptions are, in point of intellectual quality,
identical.

Imagine, then, a particle entering the film of air
where it possesses this precise thickness.   To enter the
film, its attracted end must be presented.   Within the
film it is able to turn *once* completely round; at the
other side of the film its attracted pole will be again
presented ; it will, therefore, enter the glass at the opposite side of the film *and be lost to the eye.*   All round
the place of contact, wherever the film possesses this
precise thickness, the light will equally disappear—we
shall therefore have a ring of darkness.

And now observe how well this conception falls in
with the law of proportionality discovered by Newton.
When the thickness of the film is $2\,d$, the particle has
time to perform *two* complete rotations within the
film ; when the thickness is $3\,d$, *three* complete rotations ; when $10\,d$, *ten* complete rotations are performed.   It is manifest that in each of these cases, on
arriving at the second surface of the film, the attracted
pole of the particle will be presented.   It will, there-

fore, be transmitted; and, because no light is sent to the eye, we shall have a ring of darkness at each of these places.

The bright rings follow from immediately the same conception. They occur between the dark rings, the thicknesses to which they correspond being also intermediate between those of the dark ones. Take the case of the first bright ring. The thickness of the film is $\frac{1}{2} d$; in this interval the rotating particle can perform only half a rotation. When, therefore, it reaches the second surface of the film, its repelled pole is presented; it is, therefore, driven back and reaches the eye. At all distances round the centre corresponding to this thickness the same effect is produced, and the consequence is a ring of brightness. The other bright rings are similarly accounted for. At the second one, where the thickness is $1\frac{1}{2} d$, a rotation and a half is performed; at the third, two rotations and a half; and at each of these places the particles present their repelled poles to the lower surface of the film. They are therefore sent back to the eye, and produce there the impression of brightness. This analysis, though involving difficulties when closely scrutinised, enables us to see how the theory of fits may have grown into consistency in the mind of Newton.

It has been already stated that the Emission Theory assigned a greater velocity to light in glass and water than in air or stellar space; and that on this point it was at direct issue with the theory of undulation, which makes the velocity in air or stellar space greater than in glass, or water. By an experiment proposed by Arago, and executed with consummate skill by Foucault and Fizeau, this question was brought to a crucial test, and decided in favour of the theory of undulation.

In the present instance also the two theories are at variance. Newton assumed that the action which produces the alternate bright and dark rings took place at a *single surface*; that is, the second surface of the film. The undulatory theory affirms that the rings are caused by the interference of waves reflected from both surfaces. This also has been demonstrated by experiment. By proper arrangements, as we shall afterwards learn, we may abolish reflection from one of the surfaces of the film, and when this is done the rings vanish altogether.

Rings of feeble intensity are also formed by *transmitted* light. These are referred by the undulatory theory to the interference of waves which have passed *directly* through the film, with others which have suffered *two* reflections within the film. They are thus completely accounted for.

Newton's espousal of the emission theory is said to have retarded scientific discovery. It might, however, be questioned whether, in the long run, the errors of great men have not really their effect in rendering intellectual progress rhythmical, instead of permitting it to remain uniform, the 'retardation' in each case being the prelude to a more impetuous advance. It is confusion and stagnation, rather than error, that we ought to avoid. Thus, though the undulatory theory was held back for a time, it gathered strength in the interval, and its development within the last half century has been so rapid and triumphant as to leave no rival in the field. We have now

to turn to the investigation of new classes of phenomena, of which it alone can render a satisfactory
account.

Newton, who was familiar with the idea of an ether,
and who introduced it in some of his speculations,
objected, as already stated, that if light consisted of
waves shadows could not exist; for that the waves
would bend round the edges of opaque bodies and
agitate the ether behind them. He was right in
affirming that this bending ought to occur, but wrong
in supposing that it does not occur. The bending is
real, though in all ordinary cases it is masked by the
action of interference. This inflection of the light
receives the name of *Diffraction.*

To study the phenomena of diffraction it is necessary
that our source of light should be a physical point,
or a fine line; for when luminous surfaces are employed
the waves issuing from different points of the surface
obscure and neutralize each other. A *point* of light of
high intensity is obtained by admitting the parallel rays
of the sun through an aperture in a window shutter, and
concentrating the beam by a lens of short focus. The
small solar image at the focus constitutes a suitable
point of light. The image of the sun formed on the
convex surface of a glass bead, or of a watch-glass
blackened within, though less intense, will also answer.
An intense *line* of light is obtained by admitting the
sunlight through a slit, and sending it through a
strong cylindrical lens. The slice of light is contracted to
a physical line at the focus of the lens. A glass tube
blackened within and placed in the light, reflects from
its surface a luminous line which, though less intense,
also answers the purpose.

In the experiment now to be described a vertical
slit of variable width is placed in front of the electric
lamp, and this slit is looked at from a distance through
another vertical slit, also of variable aperture, and held
in the hand.

The light of the lamp being, in the first place,
rendered monochromatic by placing a pure red glass in
front of the slit, when the eye is placed in the straight
line drawn through both slits an extraordinary appearance (shown in fig. 18) is observed.   Firstly, the slit in

Fig. 18.

front of the lamp is seen as a vivid rectangle of light;
but right and left of it is a long series of rectangles,
decreasing in vividness, and separated from each other
by intervals of absolute darkness.

The breadth of these bands is seen to vary with the
width of the slit held before the eye.   When the slit
is widened the bands become narrower, and they crowd
more closely together; when the slit is narrowed, the
individual bands widen and also retreat from each other,
leaving between them wider spaces of darkness than
before.

Leaving everything else unchanged, let a blue glass
or a solution of ammonia-sulphate of copper, which
gives a very pure blue, be placed in the path of the

light. A series of blue bands is thus obtained, exactly
like the former in all respects save one; the blue
rectangles are *narrower*, and they are *closer together*
than the red ones.

If we employ colours of intermediate refrangibilities,
which we may do by causing the different colours of a
spectrum to shine through the slit, we obtain bands of
colour intermediate in width and occupying intermediate positions between those of the red and blue. The
aspect of the bands in red, green, and violet light is
represented in fig. 19. When *white light*, therefore,

FIG. 19.

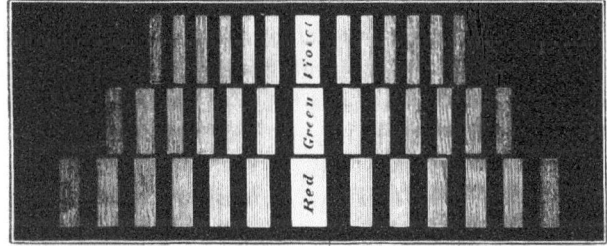

passes through the slit the various colours are not
superposed, and instead of a series of monochromatic
bands, separated from each other by intervals of darkness, we have a series of coloured spectra placed side
by side. When the distant slit is illuminated by a
candle flame, instead of the more intense electric light;
or when a distant platinum wire raised to a white heat
by an electric current is employed, substantially the
same effects are observed.

What is the meaning of these experiments, and how
are the lateral images of the slit produced? Of these
and of a multitude of similar results the Emission

Theory is incompetent to offer any satisfactory explanation. Let us see how they are accounted for by the Theory of Undulation.

And here, with the view of reaching absolute clearness, I must make an appeal to that faculty the importance of which I have dwelt upon so earnestly here and elsewhere—the faculty of imagination. Figure yourself upon the sea-shore, with a well-formed wave advancing. Take a line of particles along the front of the wave, all at the same distance below the crest; they are all rising in the same manner and at the same rate. Take a similar line of particles on the back of the wave, they are all falling in the same manner and at the same rate. Take a line of particles along the crest, they are all in the same condition as regards the motion of the wave. The same is true for a line of particles along the furrow of the wave.

The particles referred to in each of these cases respectively being in the same condition as regards the motion of the wave, are said to be in the same *phase* of vibration. But if you compare a particle on the front of the wave with one at the back; or more generally, if you compare together any two particles not occupying the same position in the wave, their conditions of motion not being the same, they are said to be in different phases of vibration. If one of the particles lie upon the crest, and the other on the furrow of the wave, then, as one is about to rise and the other about to fall, they are said to be in *opposite* phases of vibration.

There is still another point—and it is one of the utmost importance as regards our present subject—to be cleared up. Let O (fig. 20) be a point in still water which,

when disturbed, produces a series of circular waves : the
disturbance necessary to produce these waves is simply
an oscillation up and down of the point O.   Let *m n* be

FIG. 20.

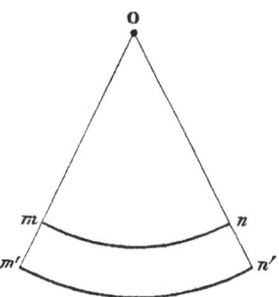

the position of the ridge of one of the waves at any moment, and *m′ n′* its position a second or two afterwards.
Now every particle of water, as the wave passes it, oscillates, as we have learned, up and down.   If, then, this
oscillation be a sufficient origin of wave motion, then
each distinct particle of the wave *m n* ought to be such
an origin and to give birth to a series of circular waves.
This is the important point up to which I wished to
lead you.   Every particle of the wave *m n does* act
in this way.   Taking each particle as a centre, and
surrounding it by a circular wave with a radius equal
to the distance between *m n* and *m′ n′*, the coalescence
of all these little waves would build up the larger
ridge *m′ n′* exactly as we find it built up in nature.
Here, in fact, we resolve the wave-motion into its
elements, and having succeeded in doing this we shall
have no great difficulty in applying our knowledge to
optical phenomena.

Now let us return to our slit, and, for the sake of simplicity, we will first consider the case of monochromatic light. Conceive a series of waves of ether advancing from the first slit towards the second, and finally filling the second slit. When each wave passes through the latter it not only pursues its direct course to the retina, but diverges right and left, tending to throw into motion the entire mass of the ether behind the slit. In fact, as already explained, *every point of the wave which fills the slit is itself a centre of a new wave-system, which is transmitted in all directions through the ether behind the slit.* This is the celebrated principle of Huyghens: we have now to examine how these secondary waves act upon each other.

Let us first regard the central band of the series. Let A P (fig. 21) be the width of the aperture held before the

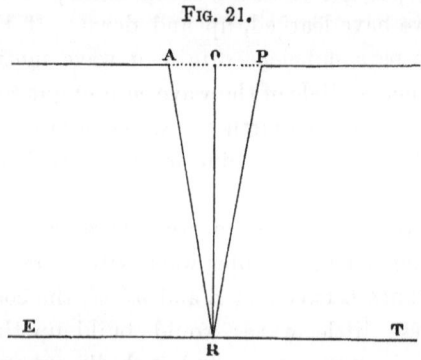

Fig. 21.

eye, grossly exaggerated of course, and let the dots across the aperture represent ether particles, all in the same phase of vibration. Let E T represent a portion of the

retina. From O, in the centre of the slit, let a perpendicular O R be imagined drawn upon the retina. The
motion communicated to the point R will then be the
sum of all the motions emanating in this direction
from the ether particles in the slit. Considering the
extreme narrowness of the aperture we may, without
sensible error, regard all points of the wave A P as
equally distant from R. No one of the partial
waves lags sensibly behind the others : hence, at R, and
in its immediate neighbourhood, we have no sensible
reduction of the light by interference. This undi-
minished light produces the brilliant central band of
the series.

Let us now consider those waves which diverge
laterally behind the slit. In this case, the waves from
the two sides of the slit have, in order to converge
upon the retina, to pass over unequal distances. Let
A P (fig. 22), represent, as before, the width of the

FIG. 22.

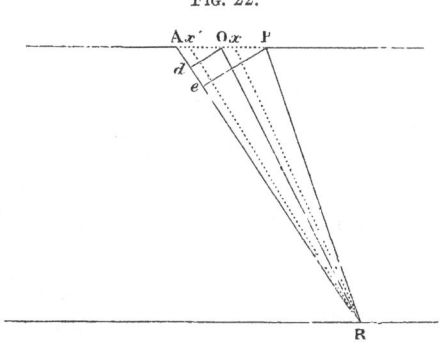

second slit. We have now to consider the action of
the various parts of the wave A P upon a point R' of

the retina, not situated in the line joining the slits.
Let us take the particular case in which the difference
in path from the two marginal points A, P, to the retina
is a whole wave-length of the red light; how must this
difference affect the final illumination of the retina?

Let us fix our attention upon the particular oblique
line that passes through the *centre* O of the slit to the
retina at R'. The difference of path between the waves
which pass along this line and those from the two
margins is, in the case here supposed, half a wave-
length. Make *e* R' equal to P R', join P and *e*, and
draw O *d* parallel to P *e*. A *e* is then the length of a
wave of light, while A *d* is half a wave-length. Now
the least reflection will make it clear that not only
is there discordance between the central and marginal
waves, but that every line of waves such as *x* R', on
the one side of O R', finds a line *x'* R' upon the other
side of O R, from which its path differs by half an
undulation, with which, therefore, it is in complete
discordance. The consequence is that the light on the
one side of the central line will completely abolish the
light on the other side of that line, absolute darkness
being the result of their coalescence. The first dark
interval of our series of bands is thus accounted for.
It is produced by an obliquity of direction which causes
the paths of the marginal waves to be *a whole wave-
length* different from each other.

When the difference between the paths of the marginal waves is *half a wave-length*, a partial destruction
of the light is effected. The luminous intensity corresponding to this obliquity is a little less than one-half
—accurately 0·4—that of the undiffracted light.

If the paths of the marginal waves be three semi-undulations different from each other, and if the whole beam be divided into three equal parts, two of these parts will, for the reasons just given, completely neutralize each other, the third only being effective. Corresponding, therefore, to an obliquity which produces a difference of three semi-undulations in the marginal waves, we have a luminous band, but one of considerably less intensity than the undiffracted central band.

With a marginal difference of path of four semi-undulations we have a second extinction of the entire beam, because here the beam can be divided into four equal parts, every two of which quench each other. A second space of absolute darkness will therefore correspond to the obliquity producing this difference. In this way we might proceed further, the general result being that, whenever the direction of wave-motion is such as to produce a marginal difference of path of an *even* number of semi-undulations, we have complete extinction ; while, when the marginal difference is an *odd* number of semi-undulations, we have only partial extinction, a portion of the beam remaining as a luminous band.

A moment's reflection will make it plain that the wider the slit the less will be the obliquity of direction needed to produce the necessary difference of path. With a wide slit, therefore, the bands, as stated, will be closer together than with a narrow one. It is also plain that the shorter the wave, the less will be the obliquity required to produce the necessary retardation. The maxima and minima of violet light must therefore fall nearer to the centre than the maxima and minima of red light. The maxima and minima of the other colours fall between these extremes. In this simple way the undulatory theory completely accounts for the extraordinary appearance above referred to.

When a slit and telescope are used, instead of the slit and naked eye, the effects are magnified and rendered more brilliant. Looking, moreover, through a properly adjusted telescope at a distant point of light with a small circular aperture in front of it, the point is seen encircled by a series of coloured bands. If monochromatic light be used, these bands are simply bright and dark, but with white light the circles display iris-colours. If a slit be shortened so as to form a square aperture, we have two series of spectra at right

Fig. 23.

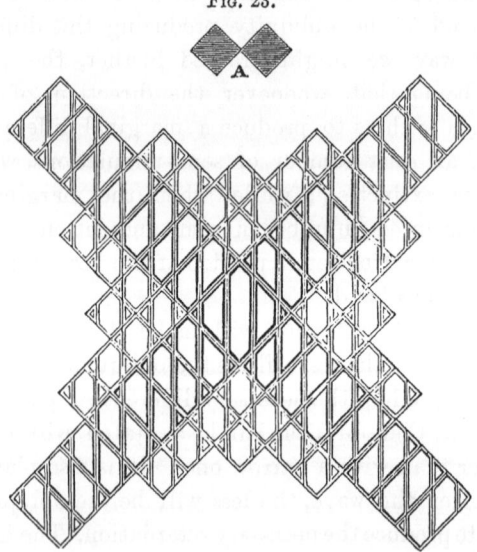

angles to each other. The effects, indeed, are capable of endless variation by varying the size, shape, and number of the apertures through which the point of light is observed. Through two square apertures, with their corners touching each other as at A, Schwerd observed the appearance shown in fig. 23. Adding two others to them, as at B, he observed the appearance represented in fig. 24. The position of every band of

FIG. 24.

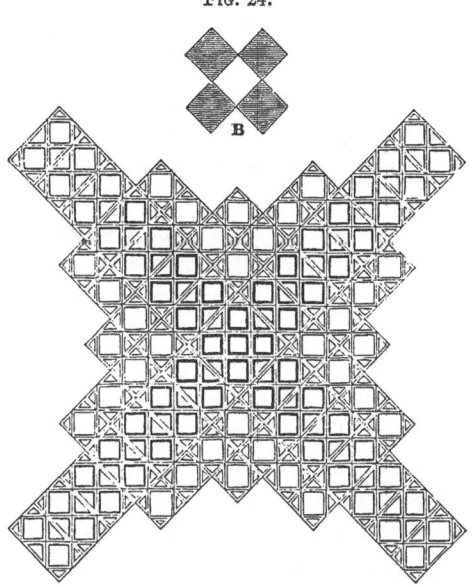

light and shade in such figures has been calculated from theory by Fresnel, Fraunhofer, Herschel, Schwerd, and others, and completely verified by experiment. Your eyes could not tell you with greater certainty of the existence of these bands than the theoretic calculation.

The street-lamps at night, looked at through the meshes of a handkerchief, show diffraction phenomena.

The diffraction effects obtained in looking through a
bird's feathers are, as shown by Schwerd, very brilliant
The iridescence of certain Alpine clouds is also an effect
of diffraction which may be imitated by the spores
of Lycopodium. When shaken over a glass plate
these spores cause a point of light, looked at through
the dusted plate, to be surrounded by coloured circles,
which rise to actual splendour when the light becomes
intense. Shaken in the air the spores produce the same
effect. The diffraction phenomena obtained during
the artificial precipitation of clouds from the vapours of
various liquids in an intensely illuminated tube are exceedingly fine.

One of the most interesting cases of diffraction by
small particles that ever came before me was that of an
artist whose vision was disturbed by vividly-coloured
circles. He was in great dread of losing his sight;
assigning as a cause of his increased fear that the
circles were becoming larger and the colours more
vivid. I ascribed the colours to minute particles in
the humours of the eye, and ventured to encourage
him by the assurance that the increase of size and
vividness on the part of the circles indicated that the
diffracting particles were becoming *smaller*, and that
they might finally be altogether absorbed. The predic-
tion was verified. It is needless to say one word on the
necessity of optical knowledge in the case of the practical oculist.

Without breaking ground on the chromatic phenomena presented by crystals, two other sources of colour
may be mentioned here. By interference in the earth's
atmosphere the light of a star, as shown by Arago, is
self-extinguished, the twinkling of the star and the

changes of colour which it undergoes being due to this
cause, Looking at such a star through an opera
glass, and shaking the glass so as to cause the
image of the star to pass rapidly over the retina,
you produce a row of coloured beads, the spaces
between which correspond to the periods of extinction.
Fine scratches drawn upon glass or polished metal
reflect the waves of light from their sides; and
some, being reflected from opposite sides of the
same furrow, interfere with and quench each other.
But the obliquity of reflection which extinguishes
the shorter waves does not extinguish the longer
ones, hence the phenomena of colour.   These are
called the colours of *striated surfaces*.   They are
beautifully illustrated by mother-of-pearl.   This shell
is composed of exceedingly thin layers, which, when cut
across by the polishing of the shell, expose their edges
and furnish the necessary small and regular grooves.
The most conclusive proof that the colours are due to
the mechanical state of the surface is to be found in
the fact, established by Brewster, that by stamping the
shell carefully upon black sealing-wax, we transfer the
grooves, and produce upon the wax the colours of
mother-of-pearl.

LECTURE III.

ONE of the objects of our last lecture, and that not the least important, was to illustrate the manner in which scientific theories are formed. They, in the first place, take their rise in the desire of the mind to penetrate to the sources of phenomena. From its infinitesimal beginnings, in ages long past, this desire has grown and strengthened into an imperious demand of man's intellectual nature. It long ago prompted Cæsar to say that he would exchange his victories for a glimpse of the sources of the Nile; it wrought itself into the atomic theories of Lucretius; it impels Darwin to those daring speculations which of late years have so agitated the public mind. But in no case in framing theories does the imagination create its materials. It expands, diminishes, moulds and refines, as the case may be, materials derived from the world of fact and observation.

This is more evidently the case in a theory like that of light, where the motions of a subsensible medium, the ether, are presented to the mind. But no theory escapes the condition. Newton took care not to encumber the idea of gravitation with unnecessary physical conceptions; but we know that he indulged in them, though he did not connect them with his theory. But even the theory as it stands did not enter the mind as a revelation dissevered from the world of experience. The germ of the conception that the sun and planets are held together by a force of attraction is to be found in the fact that a magnet had been previously seen to attract iron. The notion of matter attracting matter came thus from without, not from within. In our present lecture the magnetic force must serve us as the portal into a new subsensible domain; but in the first place we must master its elementary phenomena.

The general facts of magnetism are most simply illustrated by a magnetized bar of steel, commonly called a bar magnet. Placing such a magnet upright upon a table, and bringing a magnetic needle near its bottom, one end of the needle is observed to retreat from the magnet, while the other as promptly approaches. The needle is held quivering there by some invisible influence exerted upon it. Raising the needle along the magnet, but still avoiding contact, the rapidity of its oscillations decreases, because the force acting upon it becomes weaker. At the centre the oscillations cease. Above the centre, the end of the needle which had been previously drawn towards the magnet

retreats, and the opposite end approaches. As we ascend higher, the oscillations become more violent,
because the force becomes stronger.   At the upper end
of the magnet, as at the lower, the force reaches a
maximum ; but all the lower half of the magnet, from
E to S (fig. 25), attracts one end of the needle, while
all the upper half, from E to N, attracts the opposite
end. This *doubleness* of the magnetic force is called
polarity, and the points near the ends of the magnet in
which the forces seem concentrated are called its *poles*.

FIG. 25.

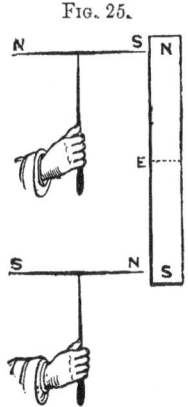

What, then, will occur if we break this magnet in
two at the centre E ?   Will each of the separate halves
act as it did when it formed part of the whole magnet?
No ; each half is in itself a perfect magnet, possessing
two poles.   This may be proved by breaking something
of less value than the magnet—the steel of a lady's
stays, for example, hardened and magnetized.   It acts
like the magnet.   When broken, each half acts like
the whole ; and when these parts are again broken, we

H

have still the perfect magnet, possessing, as in the first
instance, two poles.   Push your breaking to its utmost
sensible limit, you cannot stop there.   The bias derived
from observation will infallibly carry you beyond the
bourne of the senses, and compel you to regard this
thing that we call magnetic polarity as resident in the
ultimate particles of the steel.   You come to the
conclusion that each atom of the magnet is endowed
with this polar force.

Like all other forces, this force of magnetism is
amenable to mechanical laws ; and, knowing the direction and magnitude of the force, we can predict its
action.   Placing a small magnetic needle near a bar
magnet, it takes up a determinate position.   That
position might be deduced theoretically from the
mutual action of the poles.   Moving the needle round
the magnet, for each point of the surrounding space
there is a definite direction of the needle, and no
other.   A needle of iron will answer as well as the
magnetic needle ; for the needle of iron is magnetized
by the magnet, and acts exactly like a steel needle
independently magnetized.

If we place two or more needles of iron near the magnet, the action becomes more complex, for then the
needles are not only acted on by the magnet, but
they act upon each other.   And if we pass to smaller
masses of iron—to iron filings, for example—we find
that they act substantially as the needles, arranging
themselves in definite forms, in obedience to the magnetic action.

Placing a sheet of paper or glass over this bar
magnet and showering iron filings upon the paper, I
notice a tendency of the filings to arrange themselves

in determinate lines. They cannot freely follow this tendency, for they are hampered by the friction against the paper. They are helped by tapping the paper ; each tap releasing them for a moment, and enabling

Fig. 26.

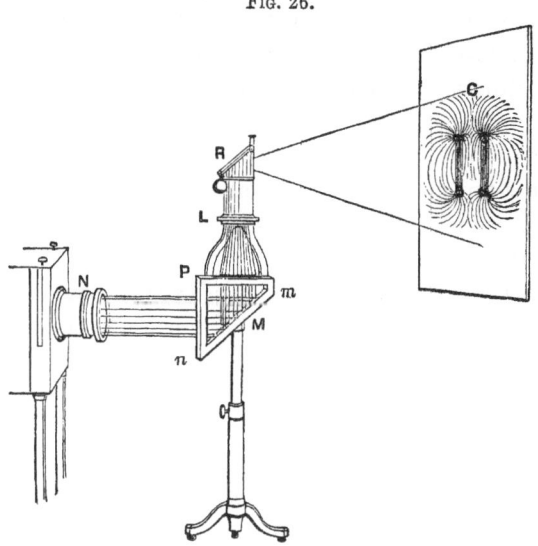

N is the nozzle of the lamp ; M a plane mirror, reflecting the beam upwards. At P the magnets and iron filings are placed ; L is a lens which forms an image of the magnets and filings ; and R is a totally-reflecting prism, which casts the image G upon the screen.

them to follow their tendencies. But this is an experiment which can only be seen by myself. To enable you all to see it, I take a pair of small magnets and by a simple optical arrangement throw the magnified images of the magnets upon the screen. Scattering iron filings over the glass plate to which the small magnets are attached, and tapping the plate, you see the arrangement of the iron filings in those magnetic curves which have been so long familiar to scientific men.[1]

The aspect of these curves so fascinated Faraday that the greater portion of his intellectual life was devoted to pondering over them. He invested the space through which they run with a kind of materiality; and the probability is that the progress of science, by connecting the phenomena of magnetism with the luminiferous ether, will prove these 'lines of force,' as Faraday loved to call them, to represent a condition of this mysterious substratum of all radiant action.

But it is not the magnetic curves, as such, but their relationship to theoretic conceptions that we have now to consider. By the action of the bar magnet upon the needle we obtain a notion of a polar force; by the breaking of the strip of magnetized steel, we attain the notion that polarity can attach itself to the ultimate particles of matter. The experiment with the iron filings introduces a new idea into the mind; the idea, namely, of *structural arrangement*. Every pair of filings possesses four poles, two of which are attractive and two repulsive. The attractive poles approach, the repulsive poles retreat; the consequence being a certain definite arrangement of the particles with reference to each other.

Now, this idea of structure, as produced by polar force, opens a way for the intellect into an entirely new region, and the reason you are asked to accompany me into this region is, that our next inquiry relates to the action of crystals upon light. Before I speak of this

[1] Very beautiful specimens of these curves have been recently obtained and *fixed* by my distinguished friend, Prof. Mayer, of Hoboken, to whom I am indebted for the original of the woodcut placed in front of this Lecture.

action, I wish you to realise intellectually the process of crystalline architecture. Look then into a granite quarry, and spend a few minutes in examining the rock. It is not of perfectly uniform texture. It is rather an agglomeration of pieces, which, on examination, present curiously-defined forms. You have there what mineralogists call quartz, you have felspar, you have mica. In a mineralogical cabinet, where these substances are preserved separately, you will obtain some notion of their forms. You will see there, also, specimens of beryl, topaz, emerald, tourmaline, heavy spar, fluor-spar, Iceland spar—possibly a full-formed diamond, as it quitted the hand of Nature, not yet having got into the hands of the lapidary.

These crystals, you will observe, are put together according to law; they are not chance productions; and, if you care to examine them more minutely, you will find their architecture capable of being to some extent revealed. They often split in certain directions before a knife-edge, exposing smooth and shining surfaces, which are called planes of cleavage; and by following these planes you sometimes reach an internal form, disguised beneath the external form of the crystal. Ponder these beautiful edifices of a hidden builder. You cannot help asking yourself how they were built; and familiar as you now are with the notion of a polar force, and the ability of that force to produce structural arrangement, your inevitable answer will be, that those crystals are built by the play of polar forces with which their molecules are endowed. In virtue of these forces, atom lays itself to atom in a perfectly definite way, the final visible form of the crystal depending upon this play of its molecules.

Everywhere in Nature we observe this tendency to

run into definite forms, and nothing is easier than to give scope to this tendency by artificial arrangements. Dissolve nitre in water, and allow the water slowly to evaporate; the nitre remains, and the solution soon becomes so concentrated that the liquid condition can no longer be preserved. The nitre-molecules approach each other, and come at length within the range of their polar forces. They arrange themselves in obedience to these forces, a minute crystal of nitre being at first produced. On this crystal the molecules continue to deposit themselves from the surrounding liquid. The crystal grows, and finally we have large prisms of nitre, each of a perfectly definite shape. Alum crystallizes with the utmost ease in this fashion. The resultant crystal is, however, different in shape from that of nitre, because the poles of the molecules are differently disposed. If they be only *nursed* with proper care, crystals of these substances may be caused to grow to a great size.

The condition of perfect crystallization is, that the crystallizing force shall act with deliberation. There should be no hurry in its operations; but every molecule ought to be permitted, without disturbance from its neighbours, to exercise its own molecular rights. If the crystallization be too sudden, the regularity disappears. Water may be saturated with sulphate of soda, dissolved when the water is hot, and afterwards permitted to cool. When cold the solution is supersaturated; that is to say, more solid matter is contained in it than corresponds to its temperature. Still the molecules show no sign of building themselves together.

This is a very remarkable, though a very common fact. The molecules in the centre of the liquid are so hampered

by the action of their neighbours that freedom to follow
their own tendencies is denied to them.   Fix your
mind's eye upon a molecule within the mass.   It wishes
to unite with its neighbour to the right, but it wishes
equally to unite with its neighbour to the left; the
one tendency neutralizes the other, and it unites with
neither.   We have here, in fact, translated into molecular action, the well-known suspension of animal
volition produced by two equally inviting bundles of
hay.   But, if a crystal of sulphate of soda be dropped
into the solution, the molecular indecision ceases.   On
the crystal the adjacent molecules will immediately
precipitate themselves; on these again others will be
precipitated, and this act of precipitation will continue
from the top of the flask to the bottom, until the
solution has, as far as possible, assumed the solid
form.   The crystals here produced are small, and confusedly arranged.   The process has been too hasty to
admit of the pure and orderly action of the crystallizing force.   It typifies the state of a nation in which
natural and healthy change is resisted, until society
becomes, as it were, supersaturated with the desire for
change, the change being then effected through confusion and revolution, which a wise foresight might
have avoided.

Let me illustrate the action of crystallizing force by
two examples of it: Nitre might be employed, but
another well-known substance enables me to make the
experiment in a better form.   The substance is common sal-ammoniac, or chloride of ammonium, dissolved
in water.   Cleansing perfectly a glass plate, the solution of the chloride is poured over the glass, to which,
when the plate is set on edge, a thin film of the liquid

adheres. Warming the glass slightly, evaporation is promoted, but by evaporation the water only is removed. The plate is then placed in a solar microscope, and an image of the film is thrown upon a white screen. The warmth of the illuminating beam adds itself to that already imparted to the glass plate, so that after a moment or two the dissolved salt can no longer exist in the liquid condition. Molecule then closes with molecule, and you have a most impressive display of crystallizing energy overspreading the whole screen. You may produce something similar if you breathe upon the frost-ferns which overspread your window-panes in winter, and then observe through a pocket lens the subsequent recongelation of the film.

In this case the crystallizing force is hampered by the adhesion of the film to the glass; nevertheless, the play of power is strikingly beautiful. Sometimes the crystals start from the edge of the film and run through it from that edge, for, the crystallization being once started, the molecules throw themselves by preference on the crystals already formed. Sometimes the crystals start from definite nuclei in the centre of the film; every small crystalline particle which rests in the film furnishing a starting-point. Throughout the process you notice one feature which is perfectly unalterable, and that is, angular magnitude. The spiculæ branch from the trunk, and from these branches others shoot; but the angles enclosed by the spiculæ are unalterable. In like manner you may find alum-crystals, quartz-crystals, and all other crystals, distorted in shape. They are thus far at the mercy of the accidents of crystallization; but in one particular they assert their superiority over all such accidents—*angular magnitude* is always rigidly preserved.

My second example of the action of crystallizing force is this : By sending a voltaic current through a liquid, you know that we decompose the liquid, and if it contains a metal, we liberate this metal by the electrolysis. This small cell contains a solution of acetate of lead, which is chosen for our present purpose because lead lends itself freely to this crystallizing power. Into the cell are dipped two very thin platinum wires, and these are connected by other wires with a small voltaic battery. On sending the voltaic current through the solution, the lead will be slowly severed from the atoms with which it is now combined ; it will be liberated upon one of the wires, and at the moment of its liberation it will obey the polar forces of its atoms, and produce crystalline forms of exquisite beauty. They are now before you, sprouting like ferns from the wire, appearing indeed like vegetable growths rendered so rapid as to be plainly visible to the naked eye. On reversing the current, these wonderful lead-fronds will dissolve, while from the other wire filaments of lead dart through the liquid. In a moment or two the growth of the lead-trees recommences, but they now cover the other wire.

In the process of crystallization, Nature first reveals herself as a builder. Where do her operations stop? Does she continue by the play of the same forces to form the vegetable, and afterwards the animal! Whatever the answer to these questions may be, trust me that the notions of the coming generations regarding this mysterious thing, which some have called ' brute matter,' will be very different from those of the generations past.

There is hardly a more beautiful and instructive example of this play of molecular force than that furnished by the case of water. You have seen the exquisite fern-like forms produced by the crystallization of a film of water on a cold window-pane.[1] You have also probably noticed the beautiful rosettes tied together by the crystallizing force during the descent of a snow-shower on a very calm day. The slopes and summits of the Alps are loaded in winter with these blossoms of the frost. They vary infinitely in detail of beauty, but the same angular magnitude is preserved throughout : an inflexible power binding spears and spiculæ to the angle of 60 degrees.

The common ice of our lakes is also ruled in its deposition by the same angle. You may sometimes see in freezing water small crystals of stellar shapes, each star consisting of six rays, with this angle of 60° between every two of them. This structure may be revealed in ordinary ice. In a sunbeam, or, failing that, in our electric beam, we have an instrument delicate enough to unlock the frozen molecules without disturbing the order of their architecture. Cutting from clear, sound, regularly-frozen ice a slab parallel to the planes of freezing, and sending a sunbeam through such a slab, it liquefies internally at special points, round each point a six-petalled liquid flower of exquisite beauty being formed. Crowds of such flowers are thus produced. From an ice-house we sometimes take blocks of ice presenting misty spaces in the otherwise continuous mass ; and when we inquire

[1] A specimen of the plumes produced by water crystallization is figured in the Frontispiece; an account of it will be found in the Appendix.

into the cause of this mistiness, we find it to be due to myriads of small six-petalled flowers, into which the ice has been resolved by the mere heat of conduction.

A moment's further devotion to the crystallization of water will be well repaid; for the sum of qualities which renders this substance fitted to play its part in Nature may well excite wonder and stimulate thought. Like almost all other substances, water is expanded by heat and contracted by cold. Let this expansion and contraction be first illustrated:

A small flask is filled with coloured water, and stopped with a cork. Through the cork passes a glass tube water-tight, the liquid standing at a certain height in the tube. The flask and its tube resemble the bulb and stem of a thermometer. Applying the heat of a spirit lamp, the water rises in the tube, and finally trickles over the top. Expansion by heat is thus illustrated.

Removing the lamp and piling a freezing mixture round the flask, the liquid column falls, thus showing the contraction of the water by the cold. But let the freezing mixture continue to act: the falling of the column continues to a certain point; it then ceases. The top of the column remains stationary for some seconds, and afterwards begins to rise. The contraction has ceased, and *expansion by cold* sets in. Let the expansion continue till the liquid trickles a second time over the top of the tube. The freezing mixture has here produced to all appearance the same effect as the flame. In the case of water, contraction by cold ceases, and expansion by cold sets in at the definite temperature of 39° Fahr. Crystallization has virtually here commenced, the molecules preparing themselves for the subsequent act of solidification which occurs at 32°, and in which the expansion suddenly culminates. In virtue of this expansion, ice, as you know, is lighter than water in the proportion of 8 to 9.[1]

A molecular problem of great interest is here involved, and I wish now to place before you, for the satisfaction of your minds, a possible solution of the problem :—

Consider, then, the ideal case of a number of magnets deprived of weight, but retaining their polar forces. If we had a mobile liquid of the specific gravity of steel, we might, by making the magnets float in it, realize this state of things, for in such a liquid the magnets would neither sink nor swim. Now, the principle of gravitation enunciated by Newton is that every particle of matter, of every kind, attracts every other particle with a force varying as the inverse square of the distance. In virtue of the attraction of gravity, then, the magnets, if perfectly free to move, would slowly approach each other.

But besides the unpolar force of gravity, which belongs to matter in general, the magnets are endowed with the polar force of magnetism. For a time, however, the polar forces do not come sensibly into play. In this condition the magnets resemble our water-molecules

---

[1] In a little volume entitled 'Forms of Water,' I have mentioned that cold iron floats upon molten iron. In company with my friend Sir William Armstrong, I had repeated opportunities of witnessing this fact in his works at Elswick, 1863. Faraday, I remember, spoke to me subsequently of the completeness of iron castings as probably due to the swelling of the metal on solidification. Beyond this, I have given the subject no special attention; and I know that many intelligent iron-founders doubt the fact of expansion. It is quite possible that the solid floats because it is not *wetted* by the molten iron, its volume being virtually augmented by capillary repulsion. Certain flies walk freely upon water in virtue of an action of this kind. With bismuth, however, it is easy to burst iron bottles by the force of solidification.

at the temperature say of 50°.   But the magnets come
at length sufficiently near each other to enable their
poles to interact.   From this point the action ceases
to be solely a general attraction of the masses.   An
attraction of special points of the masses and a repulsion of other points now come into play; and it is
easy to see that the rearrangement of the magnets consequent upon the introduction of these new forces may
be such as to require a greater amount of room. This, I
take it, is the case with our water-molecules.   Like the
magnets, they approach each other for a time *as wholes.*
Previous to reaching the temperature 39° Fahr., the
polar forces had doubtless begun to act, but it is at
this temperature that their action exactly balances the
contraction due to cold.   At lower temperatures, as
regards change of volume, the polar forces predominate.
But they carry on a struggle with the force of contraction until the freezing temperature is attained.   The
molecules then close up to form solid crystals, a considerable augmentation of volume being the immediate
consequence.

We have now to exhibit the bearings of this act of
crystallization upon optical phenomena.   According to
the undulatory theory, the velocity of light in water and
glass is less than in air.   Consider, then, a small portion of a wave issuing from a point of light so distant
that the portion may be regarded as practically plane.
Moving vertically downwards, and impinging on an
horizontal surface of glass or water, the wave would go
through the medium without change of direction.   But,
as the velocity in glass and water is less than the
velocity in air, the wave would be retarded on passing
into the denser medium.

But suppose the wave, before reaching the glass, to be *oblique* to the surface; that end of the wave which first reaches the medium will be the first retarded by it, the other portions as they enter the glass being retarded in succession. It is easy to see that this retardation of the one end of the wave must cause it to swing round and change its front, so that when the wave has fully entered the glass its course is oblique to its original direction. According to the undulatory theory, light is thus *refracted*.

With these considerations to guide us, let us follow the course of a beam of monochromatic light through our glass prism. The velocity in air is to its velocity in glass

FIG. 27.

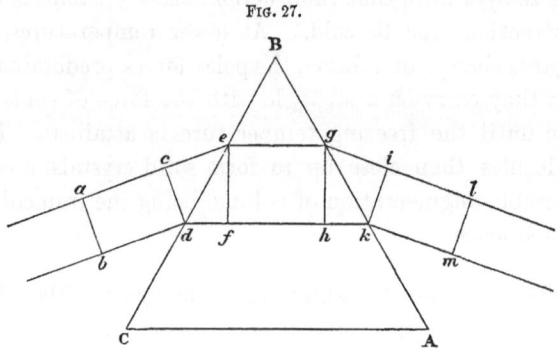

as 2 : 3. Let A B C (fig. 27) be the section of our prism, and *a b* the section of a plane wave approaching it in the direction of the arrow. When it reaches *c d*, one end of the wave is on the point of entering the glass, and while the portion of the wave still in the air passes over the distance *c e*, the wave in the glass will have passed over only two-thirds of this distance, or *d f*. The line *e f* now marks the front of the wave.

Immersed wholly in the glass it pursues its way to $g\ h$, where the end g of the wave is on the point of escaping into the air.   During the time required by the end h of the wave to pass over the distance $h\ k$ to the surface of the prism, the other end g, moving more rapidly, will have reached the point i.   The wave, therefore, has again changed its front, so that after its emergence from the prism it will pass on to $l\ m$, and subsequently in the direction of the arrow.   The refraction of the beam is thus completely accounted for; and it is, moreover, based upon actual experiment, which proves that the ratio of the velocity of light in glass to its velocity in air is that here mentioned.   It is plain that if the change of velocity on entering the glass was greater, the refractor also would be greater.

The two elements of rapidity of propagation, both of sound and light, in any substance whatever, are *elasticity* and *density*, the speed increasing with the former and diminishing with the latter.   The enormous velocity of light in stellar space is attainable because the ether is at the same time of infinitesimal density and of enormous elasticity.   Now the ether surrounds the atoms of all bodies, but it is not independent of them. In ponderable matter it acts as if its density were increased without a proportionate increase of elasticity; and this accounts for the diminished velocity of light in refracting bodies.   We here reach a point of cardinal importance.   In virtue of the crystalline architecture that we have been considering, the ether in many crystals possesses different densities in different directions; and the consequence is, that some of these media transmit light with two different velocities.   But as refraction depends wholly upon the change of velocity on entering the refracting medium, and is greatest where the change of velocity is greatest, we have in many crystals two different refractions. By such crystals a beam of light is divided into two. This effect is called *double refraction*.

In ordinary water, for example, there is nothing in the grouping of the molecules to interfere with the perfect homogeneity of the ether; but, when water crystallizes to ice, the case is different. In a plate of ice the elasticity of the ether in a direction perpendicular to the surface of freezing is different from what it is parallel to the surface of freezing; ice is, therefore, a double refracting substance. Double refraction is displayed in a particularly impressive manner by Iceland spar, which is crystallized carbonate of lime. The difference of ethereal density in two directions in this crystal is very great, the separation of the beam into the two halves being, therefore, particularly striking.

I am unwilling to quit this subject before raising it to unmistakable clearness in your minds. The vibrations of light being transversal, the elasticity concerned in the propagation of any ray is the elasticity at right angles to the direction of propagation. In Iceland spar there is one direction round which the crystalline molecules are symmetrically built. This direction is called the axis of the crystal. In consequence of this symmetry the elasticity is the same in all directions perpendicular to the axis, and hence a ray transmitted along the axis suffers no double refraction. But the elasticity along the axis is greater than the elasticity at right angles to the axis. Consider then a system of waves crossing the crystal in a direction perpendicular to the axis. Two directions of vibration are open to

such waves: the ether particles can vibrate parallel
to the axis or perpendicular to it. *They do both*, and
hence immediately divide themselves into two systems
propagated with different velocities. Double refraction
is the necessary consequence.

By means of Iceland spar cut in the proper direction,
double refraction is capable of easy illustration. Before
you is now projected an image of our carbon-points.
Causing the beam which builds the image to pass
through the spar, the single image is instantly divided
into two. Projecting (by the lens E, fig. 28) an image

Fig. 28.

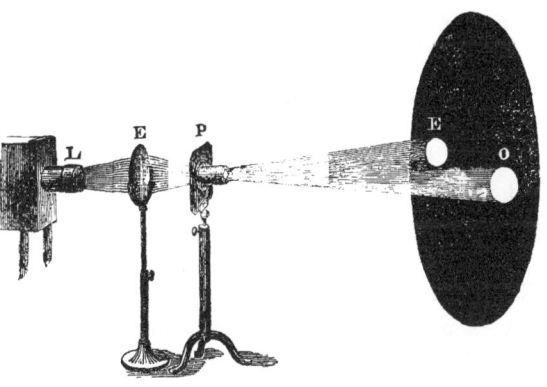

of the aperture (L) through which the light issues from
the electric lamp, and introducing the spar (P), two
luminous disks (E O) appear immediately upon the
screen instead of one.

The two beams into which the spar divides the single
incident-beam have been subjected to the closest examination. They do not behave alike. One of them
obeys the ordinary law of refraction discovered by Snell,

I

and is, therefore, called the *ordinary ray* : its index of refraction is 1·483. The other does not obey this law. Its index of refraction, for example, is not constant, but varies from a maximum of 1·654 to a minimum of 1·483 ; nor do the incident and refracted rays always lie in the same plane. It is, therefore, called the *extraordinary ray*. In calc-spar, as just stated, the ordinary ray is the most refracted. One consequence of this merits a passing notice. Pour water and bi-sulphide of carbon into two cups of the same depth, the cup that contains the more strongly-refracting liquid will appear shallower than the other. Place a piece of Iceland spar over a dot of ink ; two dots are seen, the one appearing nearer than the other to the eye. The nearest dot belongs to the most strongly-refracted ray, exactly as the nearest cup bottom belongs to the most highly refracting liquid. When you turn the spar round, the extraordinary image of the dot rotates round the ordinary one, which remains fixed. This is also the deportment of our two disks upon the screen.

The double refraction of Iceland spar was first treated in a work published by Erasmus Bartholinus, in 1669. The celebrated Huyghens sought to account for this phenomenon on the principles of the wave theory, and he succeeded in doing so. He, moreover, made highly important observations on the distinctive character of the two beams transmitted by the spar, admitting, with resigned candour, that he had not solved them, and leaving that solution to future times. Newton, reflecting on the observations of Huyghens, came to the conclusion that each of the beams transmitted by Iceland spar had two sides ; and from the analogy of this *two-sidedness* with the *two-endedness* of a magnet, wherein consists its polarity, the two beams came subsequently to be described as *polarized*.

We shall study this subject of the *polarization of light* with ease and profit by means of a crystal of tourmaline. But we must start with a clear conception of an ordinary beam of light. It has been already explained that the vibrations of the individual ether-particles are executed *across* the line of propagation. In the case of ordinary light we are to figure the ether particles as vibrating in all directions, or azimuths, as it is sometimes expressed, across this line.

Now, in the case of a plate of tourmaline cut parallel to the axis of the crystal, a beam of light incident upon the plate is divided into two, the one vibrating parallel to the axis of the crystal, the other at right angles to the axis. The grouping of the molecules, and of the ether associated with the molecules, reduces all the vibrations incident upon the crystal to these two directions. One of these beams, namely, that one whose vibrations are perpendicular to the axis, is quenched with exceeding rapidity by the tourmaline. To such vibrations many specimens of this crystal are highly opaque; so that, after having passed through a very small thickness of the tourmaline, the light emerges with all its vibrations reduced to a single plane. In this condition it is what we call a beam of *plane polarized light*.

A moment's reflection will show that, if what is here stated be correct, on placing a second plate of tourmaline with its axis parallel to the first, the light will pass through both; but that, if the axes be crossed, the light that passes through the one plate will be quenched by the other, a total interception of the light being the consequence.   Let us test this conclusion by experiment.   The image of a plate of tourmaline (*t t*, fig. 29) is now before you.   I place parallel to it another

Fig. 29.

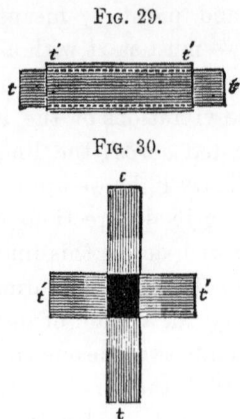

Fig. 30.

plate (*t′ t′*) : the green of the crystal is a little deepened, nothing more ; this agrees with our conclusion.   By means of an endless screw, I now turn one of the crystals gradually round, and you observe that as long as the two plates are oblique to each other, a certain portion of light gets through ; but that when they are at right angles to each other, the space common to both is a space of darkness (fig. 30).   Here also our conclusion, arrived at prior to experiment, is verified.

Let us now return to a single plate ; and here let me say that it is on the green light transmitted by the tourmaline that you are to fix your attention.   We have to illustrate the two-sidedness of that green light, in contrast to the all-sidedness of ordinary light.   The light surrounding the green image, being ordinary light, is reflected by a plane glass mirror in all directions,

the green light, on the contrary, is not so reflected.
The image of the tourmaline is now horizontal; reflected upwards, it is still green; reflected sideways,
the image is reduced to blackness, because of the in-
competency of the green light to be reflected in this
direction. Making the plate of tourmaline vertical,
and reflecting it as before, it is in the upper image that
the light is quenched; in the side image you have now
the green. This is a result of the greatest significance.
If the vibrations of light were longitudinal, like those
of sound, you could have no action of this kind; and
this very action compels us to assume that the vibrations are transversal. Picture the thing clearly. In
the one case the mirror receives the impact of the
edges of the waves, the green light being then quenched.
In the other case the *sides* of the waves strike the mirror, and the green light is reflected. To render the
extinction complete, the light must be received upon
the mirror at a special angle. What this angle is we
shall learn presently.

The quality of two-sidedness conferred upon light
by crystals may also be conferred upon it by ordinary
reflection. Malus made this discovery in 1808, while
looking through Iceland spar at the light of the sun
reflected from the windows of the Luxembourg palace
in Paris. I receive upon a plate of window-glass the
beam from our lamp; a great portion of the light reflected from the glass is polarized. The vibrations of
this reflected beam are executed, for the most part,
parallel to the surface of the glass, and when the glass
is held so that the beam shall make an angle of 58°
with the perpendicular to the glass, the *whole* of the
reflected beam is polarized. It was at this angle that

the image of the tourmaline was completely quenched in our former experiment. It is called *the polarizing angle.*

Sir David Brewster proved the angle of polarization of a medium to be that particular angle at which the refracted and reflected rays inclose a right angle.[1] The polarizing angle augments with the index of refraction. For water it is $52\frac{1}{2}°$; for glass, as already stated, $58°$; while for diamond it is $68°$.

And now let us try to make substantially the experiment of Malus. The beam from the lamp is received upon this plate of glass and reflected through the spar. Instead of two images, you see but one. So that the light, when polarized, as it now is, can only get through the spar in one direction, and consequently produce but one image. Why is this? In the Iceland spar, as in the tourmaline, all the vibrations of the ordinary light are reduced to two planes at right angles to each other; but, unlike the tourmaline, both beams are transmitted with equal facility by the spar. The two beams, in short, emergent from the spar, are polarized, their directions of vibration being at right angles to each other. When, therefore, the light was polarized by reflection, the direction of vibration in the spar which corresponded to the direction of vibration of the polarized beam transmitted it, and that direction only. Only one image, therefore, was possible under the conditions.

[1] This beautiful law is usually thus expressed: *The index of refraction of any substance is the tangent of its polarizing angle.* With the aid of this law and apparatus similar to that figured at page 15, we can readily determine the index of refracting any liquid. The refracted and reflected beams being visible, they can be caused to enclose a right angle The polarizing angle of the liquid may be thus found with the sharpest precision. It is then only necessary to seek out its natural tangent to obtain the index of refraction.

You will now observe that such logic as connects
our experiments is simply a transcript of the logic of
Nature.   On the screen before you are two disks of
light produced by the double refraction of Iceland spar.   They are, as you know, two images of the
aperture through which the light issues from the
camera.   Placing the tourmaline in front of the aperture, two images of the crystal will also be obtained;
but now let us reason out what is to be expected
from this experiment.   The light emergent from the
tourmaline is polarized.   Placing the crystal with its
axis horizontal, the vibrations of its transmitted light
will be horizontal.   Now the spar, as already stated,
has two directions of vibration, one of which at the
present moment is vertical, the other horizontal.   What
are we to conclude?   That the green light will be
transmitted along the latter, which is parallel to the
axis of the tourmaline, and not along the former, which
is perpendicular to that axis.   Hence we may infer
that one image of the tourmaline will show the ordinary green light of the crystal, while the other image
will be black.   Tested by experiment, our reasoning is
verified to the letter (fig. 31).

FIG. 31.

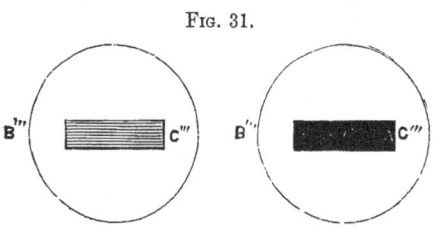

Let us push our test still further.   By means of an
endless screw, the crystal can be turned ninety degrees

round. The black image, as I turn, becomes gradually brighter, and the bright one gradually darker; at an angle of forty-five degrees both images are equally

Fig. 32.

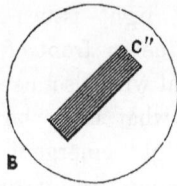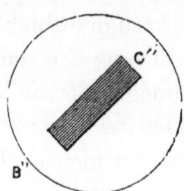

bright (fig. 32); while, when ninety degrees have been obtained, the axis of the crystal being then vertical, the bright and black images have changed places, exactly as reasoning would have led us to suppose (fig. 33.)

Given the two beams transmitted through Iceland spar, it is perfectly manifest that we have it in our power to determine instantly, by means of a plate of

Fig. 33.

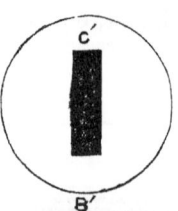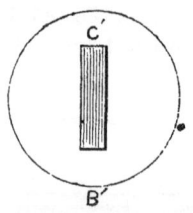

tourmaline, the directions in which the ether-particles vibrate in the two beams. The double refracting spar might be placed in any position whatever. A minute's trial with the tourmaline would enable you to determine the position which yields a black and a bright image, and from this you would at once infer the directions of vibration.

Let us reason still further together.   The two beams

FIG. 34.

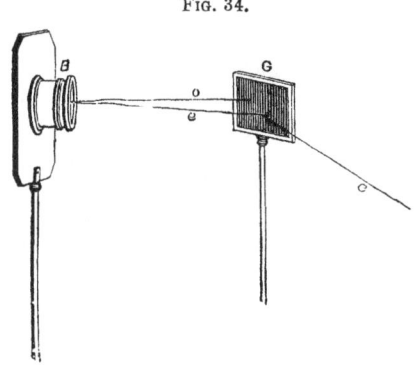

(B is the birefracting spar, dividing the incident light into the two beams *o* and *e*, G is the mirror.) The beam is here reflected *laterally*. When the reflection is *upwards*, the other beam is reflected as shown in fig. 35.

from the spar being thus polarized, it is plain that if they be suitably received upon a plate of glass at the

FIG. 35.

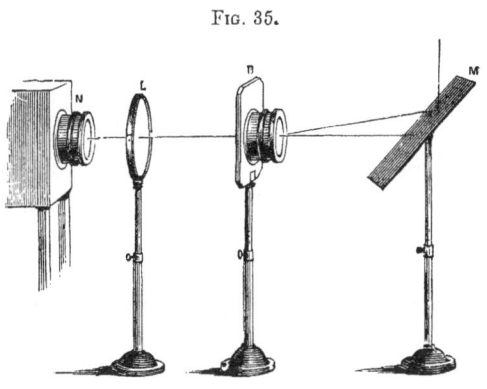

polarizing angle, one of them will be reflected, the other not. This is a simple inference from our previous knowledge ; but you observe that the inference is justified by experiment. (Figs. 34 and 35.)

I have said that the whole of the beam reflected from glass at the polarizing angle is polarized ; a word must now be added regarding the far larger portion of the light which is *transmitted* by the glass. The transmitted beam contains a quantity of polarized light equal to that of the reflected beam : but this quantity is only a fraction of the whole transmitted light. By taking two plates of glass instead of one, we augment the quantity of the transmitted polarized light ; and by taking *a bundle* of plates, we so increase the quantity as to render the transmitted beam, for all practical purposes, *perfectly* polarized. Indeed, bundles of glass plates are often employed as a means of furnishing polarized light. Interposing such a bundle at the proper angle into the paths of the two beams emergent from Iceland spar, that which, in the last experiment, failed to be reflected, is here transmitted. The plane of vibration of this transmitted light is, however, at right angles to that of the reflected light.

One word more. When the tourmalines are crossed, the space where they cross each other is black. But we have seen that the least obliquity on the part of the crystals permits light to get through both. Now suppose, when the two plates are crossed, that we interpose a third plate of tourmaline between them, with its axis oblique to both. A portion of the light transmitted by the first plate will get through this intermediate one. But, after it has got through, *its plane of vibration is changed*: it is no longer perpendicular to the axis of the crystal in front.   Hence it will get through that crystal.   Thus, by reasoning, we infer that the interposition of a third plate of tourmaline will in part abolish the darkness produced by the perpendicular crossing of the other two plates.   I have not a third plate of tourmaline; but the talc or mica which you employ in your stoves is a more convenient substance, which acts in the same way.   Between the crossed tourmalines, I introduce a film of this crystal.   You see the edge of the film slowly descending, and as it descends between the tourmalines, light takes the place of darkness.   The darkness, in fact, seems scraped away, as if it were something material.   This effect has been called, naturally but improperly, *depolarization*.   Its proper meaning will be disclosed in our next lecture.

LECTURE IV.

CHROMATIC PHENOMENA PRODUCED BY CRYSTALS ON POLARIZED LIGHT—
THE NICOL PRISM—POLARIZER AND ANALYZER—ACTION OF THICK AND
THIN PLATES OF SELENITE—COLOURS DEPENDENT ON THICKNESS—
RESOLUTION OF POLARIZED BEAM INTO TWO OTHERS BY THE SELENITE
—ONE OF THEM MORE RETARDED THAN THE OTHER—RECOMPOUNDING
OF THE TWO SYSTEMS OF WAVES BY THE ANALYZER—INTERFERENCE
THUS RENDERED POSSIBLE—CONSEQUENT PRODUCTION OF COLOURS—
ACTION OF BODIES MECHANICALLY STRAINED OR PRESSED—ACTION OF
SONOROUS VIBRATIONS—ACTION OF GLASS STRAINED OR PRESSED BY
HEAT—CIRCULAR POLARIZATION—CHROMATIC PHENOMENA PRODUCED
BY QUARTZ—THE MAGNETIZATION OF LIGHT—RINGS SURROUNDING THE
AXES OF CRYSTALS—BIAXAL AND UNIAXAL CRYSTALS—GRASP OF THE
UNDULATORY THEORY—THE COLOUR AND POLARIZATION OF SKY-LIGHT
—GENERATION OF ARTIFICIAL SKIES.

WE now stand upon the threshold of a new and
splendid optical domain.   We have penetrated as far,
perhaps, as it is at present possible to penetrate into
the arcana of crystallization.   By dwelling upon the
very phenomena which suggested it, we have mastered
the conception of plane-polarized light.   But I am
here reminded of an argument which Protestants sometimes urge against Catholics. ' You prove,' say they,
' the authenticity of the Scriptures by the authority of the
Church, and then deduce the authority of the Church
from the Scriptures : that looks like arguing in a circle.'
I may seem to lay myself open to a similar reproach
when I say that the conception of polarized light is
based upon the facts of observation, and then immediately proceed to deduce the facts of observation from
the conception of polarized light.

This objection would fairly apply if the theoretic conception were limited by the facts in which it originated. Some theories are open to this criticism, but in so far as they are so, they lack the characteristics of a true theory. The scientific intellect resembles a lamp, which does not burn and shine until ignited by the match of observation or experiment. But the light emitted after ignition may, in virtue of the mind's inherent energy, transcend a million-fold that of the match which started it. In fact, it may be said that they stand to each other in an incommensurable relation ; a few bounded and solitary facts, by their action on the mind, sufficing to liberate principles of indefinite applicability and extension.

We have this evening to illustrate and examine the chromatic phenomena produced by the action of crystals, and double-refracting bodies generally, upon polarized light, and to apply the Undulatory Theory to their elucidation. For a long time investigators were compelled to employ plates of tourmaline for this purpose, and the progress they made with so defective a means of inquiry is astonishing. But these men had their hearts in their work, and were on this account enabled to extract great results from small instrumental appliances. But for educational purposes we need far larger apparatus, and, happily, in these later times this need has been to a great extent satisfied. We have seen and examined the two beams emergent from Iceland spar, and have proved them to be polarized. If, at the sacrifice of half the light, we could abolish one of these, the other would place at our disposal a beam of polarized light incomparably stronger than any attainable from tourmaline.

The beams, as you know, are refracted differently, and from this we are able to infer that under certain circumstances the one may be totally reflected, and the other not. An able optician, named Nicol, taking advantage of this, cut a crystal of Iceland spar in two halves in a certain direction. He polished the severed surfaces, and reunited them by Canada balsam, the surface of union being so inclined to the beam traversing the spar that the ordinary ray, which is the most highly refracted, was totally reflected by the balsam, while the extraordinary ray was permitted to pass on.

Let $b\,x$, $c\,y$ (fig. 36) represent the section of an elongated rhomb of Iceland spar cloven from the crystal. Let

Fig. 36.

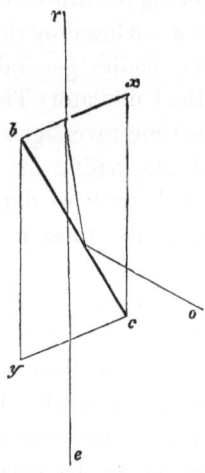

this rhomb be cut along the line $b\,c$; and the two severed surfaces, after having been polished, reunited by Canada balsam. We learned, in our first lecture,

that total reflection only takes place when a ray seeks
to escape from a more refracting to a less refracting
medium, and that it always, under these circumstances,
takes place when the obliquity is sufficient.   Now the
refractive index of Iceland spar is, for the extraordinary
ray less, and for the ordinary greater, than for Canada
balsam.   Hence, in passing from the spar to the balsam,
the extraordinary ray passes from a less refracting to
a more refracting medium, where total reflection cannot
occur; while the ordinary ray passes from a more
refracting to a less refracting medium, where total
reflection can occur.   The requisite obliquity is secured
by making the rhomb of such a length that the plane
of which $b\ c$ is the section shall be perpendicular, or
nearly so, to the two end surfaces of the rhomb $b\ x, c\ y$.

The invention of the Nicol prism was a great step in
practical optics, and quite recently such prisms have
been constructed of a size which enables audiences like
the present to witness the chromatic phenomena of
polarized light to a degree altogether unattainable a
short time ago.   The two prisms here before you belong to my excellent friend Mr. William Spottiswoode,
and they were manufactured by Mr. Ladd.   I have
with me another pair of very noble prisms, still larger
than these, manufactured for me by Mr. Browning,
who has gained so high and well-merited a reputation
in the construction of spectroscopes.*

These two Nicol prisms play the same part as the
two plates of tourmaline.   Placed with their directions
of vibration parallel, the light passes through both;

* The largest and purest prism hitherto made has been recently constructed for Mr. Spottiswoode by Messrs. Tisley & Spiller.

while when these directions are crossed the light is quenched. Introducing a film of mica between the prisms, the light is restored. But notice, when the film of mica is *thin* you have sometimes not only light, but *coloured* light. Our work for some time to come will consist of the examination of such colours. With this view, I will take a representative crystal, one easily dealt with, because it cleaves with great facility—the crystal gypsum, or selenite, or crystallized sulphate of lime. Between the crossed Nicols I place a thick plate of this crystal ; like the mica, it restores the light, but it produces no colour. With my penknife I take a thin splinter from this crystal and place it between the prisms ; the image of the splinter glows with the richest colours. Turning the prism in front, these colours gradually fade and disappear, but, by continuing the rotation until the vibrating sections of the prisms are parallel to each other, vivid colours again arise, but these colours are complementary to the former ones.

Some patches of the splinter appear of one colour, some of another. These differences are due to the different thicknesses of the film. As in the case of Hooke's thin plates, if the thickness be uniform, the colour is uniform. Here, for instance, is a stellar shape, every lozenge of the star being a film of gypsum of uniform thickness : each lozenge, you observe, shows a brilliant and uniform colour. It is easy, by shaping our films so as to represent flowers or other objects, to exhibit such objects in hues unattainable by art. Here, for example, is a specimen of heart's-ease, the colours of which you might safely defy the artist to reproduce. By turning the front Nicol 90 degrees round, we pass through a colourless phase to a series of colours complementary to the former ones. This change is still more strikingly represented by a rose-tree, which is now presented in its natural hues—a red flower and green leaves; turning the prism 90 degrees round, we obtain a green flower and red leaves. All these wonderful chromatic effects have definite mechanical causes in the motions of the ether. The principle of interference duly applied and interpreted explains them all.

By this time you have learned that the word 'light' may be used in two different senses; it may mean the impression made upon consciousness, or it may mean the physical agent which makes the impression. It is with the agent that we have to occupy ourselves at present. That agent is a substance which fills all space, and surrounds the atoms and molecules of bodies. To this interstellar and interatomic medium definite mechanical properties are ascribed, and we deal with it in our reasonings and calculations as a body possessed of these properties. In mechanics we have the composition and resolution of forces and of motions, extending to the composition and resolution of *vibrations*. We treat the luminiferous ether on mechanical principles, and, from the composition, resolution, and interference of its vibrations we deduce all the phenomena displayed by crystals in polarized light.

Let us take, as an example, the crystal of tourmaline, with which we are now so familiar. Let a vibration cross this crystal oblique to its axis. Experiment has assured us that a portion of the light will pass through. How much passes we determine in this way. Let A B, fig. 37, be the axis of the tourmaline, and, let $a\,b$ represent the amplitude of the ethereal vibration before it

K

reaches A B.   From a and b let the two perpendiculars $a\,c$ and $b\,d$ be drawn upon the axis: then $c\,d$ will be the amplitude of the transmitted vibration.

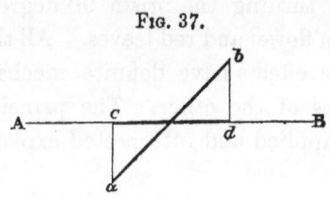

FIG. 37.

In a moment I will ask you to follow me while I endeavour to explain to you the effect observed when a film of gypsum is placed between the two Nicol's prisms. But, at the outset, it will be desirable to establish still further the analogy between the action of the prisms and that of two plates of tourmaline.   The magnified image of these plates, with their axes at right-angles to each other, is now before you.   I introduce between them a film of Selenite, and you see that by turning the film round it may be placed in a position where it has no power to abolish the darkness of the superposed portions of the tourmalines.   Why is this?   The answer is, that in the gypsum there are two directions, at right angles to each other, in which alone vibrations can take place, and that in our present experiment one of these directions is parallel to one of the axes of the tourmaline, and the other parallel to the other axis. When this is the case, the film exercises no sensible action upon the light.   But now I turn the film so as to render its directions of vibration *oblique* to the two axes; then you see it has the power, demonstrated in the last lecture, of restoring the light.

Let us now mount our Nicol's prisms, and cross them

as we crossed the tourmalines. Introducing our film of gypsum between them, you notice that in one particular position the film has no power whatever over the field of view. But, when the film is turned a little way round, the light passes. We have now to understand the mechanism by which this is effected.

Firstly, then, we have this first prism which receives the light from the electric lamp, and which is called the *polarizer*. Then we have the plate of gypsum (supposed to be placed at S, fig. 38), and then the prism in front, which is called the *analyzer*. On its emergence from the first prism, the light is polarized;

FIG. 38.

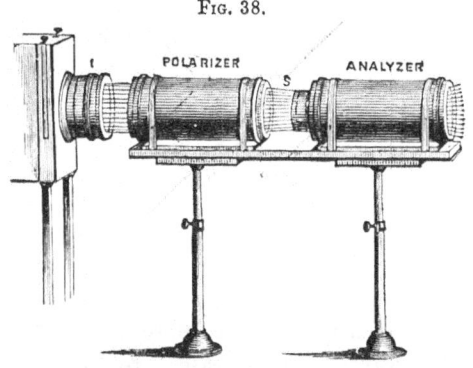

and, in the particular case now before us, its vibrations are executed in a horizontal plane. The two directions of vibration of the gypsum, placed at S, are now oblique to the horizon. Draw a rectangular cross (A B, C D fig. 39) to represent the two directions of vibration within the gypsum. Draw a line (*a b*) to represent the amplitude of the vibration from the first Nicol, when it reaches the gypsum. Let fall from the two ends of this line two perpendiculars on each of the

arms of the cross; then the distances (*c d, e f,*) between the feet of these perpendiculars represent the
amplitudes of two rectangular vibrations, *which are the
components of the first single vibration.*   Thus the
polarized ray, when it enters the gypsum, is resolved
into its two equivalents, which vibrate at right angles
to each other.

Now, in one of those rectangular directions of vibration the ether within the gypsum is more sluggish
than in the other; and, as a consequence, the waves

Fig. 39.

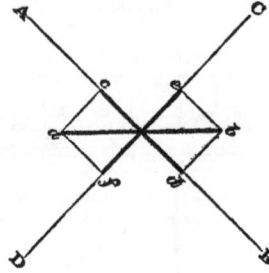

that follow this direction are more retarded than the
others. The waves in both cases are shortened when
they enter the gypsum, but the waves of the one system
are more shortened than those of the other. You can
readily imagine that in this way the one system of
waves may get half a wave-length, or indeed any number of half-wave lengths, in advance of the other. The
possibility of interference here at once flashes upon the
mind. A little consideration, however, will render it
evident that, as long as the vibrations are executed at
right angles to each other, they cannot quench each

other, no matter what the retardation may be. This
brings us at once to the part played by the analyzer,
the sole function of which is to recompound the two
vibrations emergent from the gypsum. It reduces
them to a single plane, where, if one of them be retarded
by the proper amount, extinction will occur.

But here, as in the case of thin films, the different
lengths of the waves of light come into play. Red will
require a greater thickness to produce the retardation
necessary for extinction than blue; consequently, when
the longer waves have been withdrawn by interference,
the shorter ones remain, the film of gypsum shining
with the colours which they confer. Conversely, when
the shorter waves have been withdrawn, the thickness
is such that the longer waves remain. An elementary
consideration suffices to show that, when the directions
of vibration of prisms and gypsum enclose an angle of
forty-five degrees, the colours are at their maximum
brilliancy. When the film is turned from this direction, the colours gradually fade, until, at the point
where the directions are parallel, they disappear
altogether.

Perhaps the best way of obtaining a knowledge of
these phenomena is to construct a model of thin wood or
pasteboard, representing the plate of gypsum, its planes
of vibration, and also those of the polarizer and analyzer. Two parallel pieces of the board are to be
separated by an interval which shall represent the
thickness of the film of gypsum. Between them, two
other pieces, intersecting each other at a right angle,
are to represent the planes of vibration within the film;
while attached to the two parallel surfaces outside are
two other pieces of board to represent the planes of

vibration of the polarizer and analyzer. On the two
intersecting planes the waves are to be drawn, showing
the resolution of the first polarized beam into two
others, and then the subsequent reduction of the two
systems of vibrations to a common plane by the analyser.
Following out rigidly the interaction of the two systems of waves, we are taught by such a model that all
the phenomena of colour obtained by the combination
of the waves when the planes of vibration of the two
Nicols are parallel are displaced by the *complementary*
phenomena when the Nicols are perpendicular to each
other.

In considering the next point, we will operate, for
the sake of simplicity, with monochromatic light—with
red light, for example, which is most easily obtained
pure by absorption. Supposing a certain thickness
of the gypsum, produces a retardation of half a wave-
length, twice this thickness will produce a retardation
of two half wave-lengths, three times this thickness a
retardation of three half-wave lengths, and so on.
Now, when the Nicols are parallel, the retardation of
half a wave-length, or of any *odd* number of half wave-
lengths, produces extinction; at all thicknesses, on the
other hand, which correspond to a retardation of an
even number of half wave-lengths, the two beams support each other, when they are brought to a common
plane by the analyzer. Supposing, then, that we take
a plate of a wedge-form, which grows gradually thicker
from edge to back, we ought to expect in red light a
series of recurrent bands of light and darkness; the
dark bands occurring at thicknesses which produce
retardations of one, three, five, etc., half wave-lengths,
while the bright bands occur between the dark ones.

Experiment proves the wedge-shaped film to show these bands. They are also beautifully shown by a circular film, so worked as to be thinnest at the centre, and gradually increasing in thickness from the centre outwards. A splendid series of rings of light and darkness is thus produced.

When, instead of employing red light, we employ blue, the rings are also seen : but, as they occur at thinner portions of the film, they are smaller than the rings obtained with the red light. The consequence of employing *white* light may be now inferred : inasmuch as the red and the blue fall in different places, we have *iris-coloured* rings produced by the white light.

Some of the chromatic effects of irregular crystallization are beautiful in the extreme. Could I introduce between our Nicols a pane of glass covered by those frost-ferns which the cold weather renders now so frequent, rich colours would be the result. The beautiful effects of the irregular crystallization of tartaric acid and other substances on glass plates, now presented to you, illustrate what you might expect from the frosted window-pane. And not only do crystalline bodies act thus upon light, but almost all bodies that possess a definite structure do the same. As a general rule, organic bodies act in this way; for their architecture implies an arrangement of the molecules and of the ether, which involves double refraction. A film of horn, or the section of a shell, for example, yields very beautiful colours in polarized light. In a tree, the ether certainly possesses different degrees of elasticity along and across the fibre ; and, were wood transparent, this peculiarity of molecular structure

would infallibly reveal itself by chromatic phenomena
like those that you have seen. But not only do
natural bodies behave in this way, but it is possible, as
shown by Brewster, to confer, by artificial strain or
pressure, a temporary double-refracting structure upon
non-crystalline bodies, such as common glass.

This is a point worthy of illustration. When I place
a bar of wood across my knee and seek to break it,
what is the mechanical condition of the bar? It
bends, and its convex surface is *strained* longitudinally ;
its concave surface, that next my knee, is longitudinally *pressed*. Both in the strained portion and in the
pressed portion the ether is thrown into a condition
which would render the wood, were it transparent,
double-refracting. For, in cases like the present, the
drawing of the molecules asunder longitudinally is
always accompanied by their approach to each other
laterally ; while the longitudinal squeezing is accompanied by lateral retreat. Each half of the bar exhibits
this antithesis, and is therefore double-refracting.

Let us now repeat the experiment with a bar of
glass. Between the crossed Nicols I introduce such a
bar. By the dim residue of light lingering upon the
screen, you see the image of the glass, but it has no
effect upon the light. I simply bend the glass bar
with my finger and thumb, keeping its length oblique
to the directions of vibration in the Nicols. Instantly
light flashes out upon the screen. The two sides of
the bar are illuminated, the edges most, for here the
strain and pressure are greatest. In passing from
longitudinal strain to longitudinal pressure, we cross a
portion of the glass where neither is exerted. This is
the so-called neutral axis of the bar of glass, and along

it you see a dark band, indicating that the glass along this axis exercises no action upon the light. By employing the force of a press, instead of the force of my finger and thumb, the brilliancy of the light is greatly augmented.

Again, I have here a square of glass which can be inserted into a press of another kind. Introducing the uncompressed square between the prisms, its neutrality is declared; but it can hardly be held sufficiently loosely in the press to prevent its action from manifesting itself. Already, though the pressure is infinitesimal, you see spots of light at the points where the press is in contact with the glass. I now turn this screw. Instantly the image of the square of glass flashes out upon the screen, and luminous spaces are seen separated from each other by dark bands.

Every pair of the adjacent luminous spaces is in opposite mechanical conditions. On one side of the dark band we have strain, on the other side pressure; while the dark band marks the neutral axis between both. I now tighten the vice, and you see colour; tighten still more, and the colours appear as rich as those presented by crystals. Releasing the vice, the colours suddenly vanish; tightening suddenly, they reappear. From the colours of a soap-bubble Newton was able to infer the thickness of the bubble, thus uniting by the bond of thought apparently incongruous things. From the colours here presented to you, the magnitude of the pressure employed might be inferred. Indeed, the late M. Wertheim, of Paris, invented an instrument for the determination of strains and pressures by the colours of polarized light, which exceeded in accuracy all previous instruments of the kind.

You know that bodies are expanded by heat and contracted by cold. If the heat be applied with perfect uniformity, no local strains or pressures come into play; but, if one portion of a solid be heated and others not, the expansion of the heated portion introduces strains and pressures which reveal themselves under the scrutiny of polarized light. When a square of common window-glass is placed between the Nicols, you see its dim outline, but it exerts no action on the polarized light. Held for a moment over the flame of a spirit-lamp, on reintroducing it between the Nicols, light flashes out upon the screen. Here, as in the case of mechanical action, you have luminous spaces of strain divided by dark neutral axes from spaces of pressure.

Let us apply the heat more symmetrically. This small square of glass is perforated at the centre, and into the orifice a bit of copper wire is introduced. Placing the square between the prisms, and heating the wire, the heat passes by conduction to the glass, through which it spreads from the centre outwards. You then see a dim cross bounding four luminous quadrants growing up and becoming gradually black by comparison with the adjacent brightness. And as, in the case of pressure, we produced colours, so here also, by the proper application of heat, gorgeous chromatic effects may be produced. The condition necessary to the production of these colours may be rendered permanent by first heating the glass sufficiently, and then cooling it, so that the chilled mass shall remain in a state of strain and pressure. Two or three examples will illustrate this point. Figs. 40 and 41 represent the figures obtained with two pieces of glass thus prepared.   Two rectangular pieces of unannealed glass, crossed and placed between the polarizer and

FIG. 40.   FIG. 41.

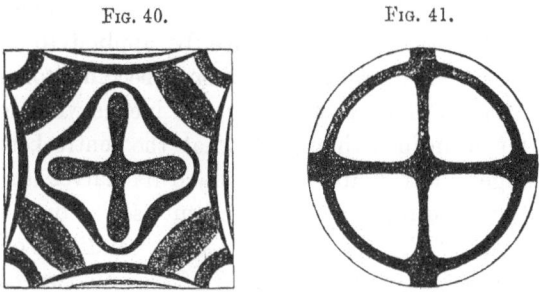

analyser, exhibit the beautiful iris fringes represented in fig. 42.

FIG. 42.

And now we have to push these considerations to a final illustration. Polarized light may be turned to account in various ways as an analyzer of molecular condition. It may, for instance, be applied to reveal the condition of a solid body when it becomes sonorous. This strip of glass six feet long, two inches wide, and a quarter of an inch thick, is held at the centre between the finger and thumb. Over one of its halves is swept a wet woollen rag; you hear an acute sound due to the vibrations of the glass. What is the condition of the glass while the sound is heard? This: its two halves lengthen and shorten in quick succession. Its two ends, therefore, are in a state of quick vibration; but at the centre the pulses from the two ends alternately meet and retreat. Between their opposing actions, the glass at the centre is kept motionless; but, on the other hand, it is alternately strained and compressed. The state of the glass may be illustrated by a row of spots of light, as the propagation of a sonorous pulse was illustrated in our second lecture. By a simple mechanical contrivance the spots are made to vibrate to and fro: the terminal dots have the largest amplitude of vibration, while those at the centre are alternately crowded together and drawn asunder, the centre one not moving at all. (In fig. 43, A B represents the glass rectangle with its centre condensed; while A′ B′ represents the same rectangle with its centre rarefied. The ends of the strip suffer neither condensation nor rarefaction.)

If we introduce the strip of glass (s s′ fig. 44) between the crossed Nicols, taking care to keep it oblique to the directions of vibration of the Nicols, and sweep our wet rubber over the glass, this may be expected to occur:

At every moment of compression the light will flash

Fig. 43.

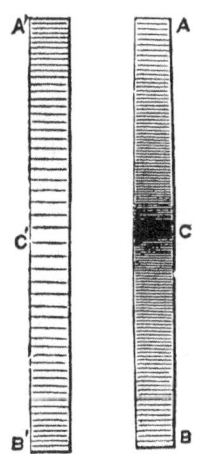

through ; at every moment of strain the light will also

Fig. 44.

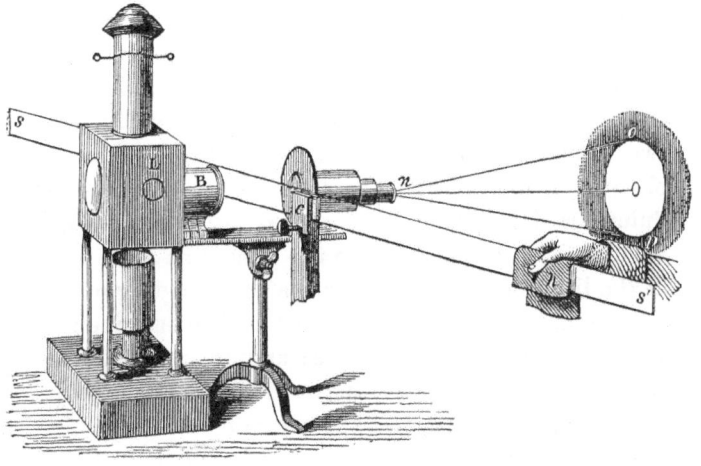

flash through ; and these states of strain and pressure will follow each other so rapidly that we may expect a permanent luminous impression to be made upon the eye. By pure reasoning, therefore, we reach the conclusion that the light will be revived whenever the glass is sounded. That it is so, experiment testifies: at every sweep of the rubber, a fine luminous disk (o) flashes out upon the screen. The experiment may be varied in this way : Placing in front of the polarizer a plate of unannealed glass, you have a series of beautifully coloured rings, intersected by a black cross. Every sweep of the rubber not only abolishes the rings, but introduces complementary ones, the black cross being for the moment supplanted by a white one. This is a modification of a beautiful experiment which we owe to Biot. His apparatus, however, confined the observation of it to a single person at a time.

But we have to follow the ether still further into its hiding-places. Suspended before you is a pendulum, which, when drawn aside and liberated, oscillates to and fro. If, when the pendulum is passing the middle point of its excursion, I impart a shock to it tending to drive it at right angles to its present course, what occurs ? The two impulses compound themselves to a vibration oblique in direction to the former one, but the pendulum still oscillates in *a plane*. But, if the rectangular shock be imparted to the pendulum when it is at the limit of its swing, then the compounding of the two impulses causes the suspended ball to describe not a straight line, but an ellipse ; and, if the shock be competent of itself to produce a vibration of the same amplitude as the first one, the ellipse becomes a circle

Why do I dwell upon these things ? Simply to make

known to you the resemblance of these gross mechanical
vibrations to the vibrations of light. I hold in my
hand a plate of quartz cut from the crystal perpendicular to its axis. This crystal thus cut possesses the
extraordinary power of twisting the plane of vibration
of a polarized ray to an extent dependent on the thickness of the crystal. And the more refrangible the
light the greater is the amount of twisting, so that,
when white light is employed, its constituent colours
are thus drawn asunder. Placing the quartz between
the polarizer and analyzer, you see this splendid red,
and, turning the analyzer in front, from right to left,
the other colours of the spectrum appear in succession.
Specimens of quartz have been found which require
the analyzer to be turned from left to right to obtain
the same succession of colours. Crystals of the first
class are therefore called right-handed, and of the
second class, left-handed crystals.

With profound sagacity, Fresnel, to whose genius
we mainly owe the expansion and final triumph of the
undulatory theory of light, reproduced mentally the
mechanism of these crystals, and showed their action
to be due to the circumstance that, in them, the waves of
ether so act upon each other as to produce the condition
represented by our rotating pendulum. Instead of
being plane polarized, the light in rock crystal is *circularly polarized*. Two such rays, transmitted along
the axis of the crystal, and rotating in opposite directions, when brought to interference by the analyzer,
are demonstrably competent to produce all the observed
phenomena.

I now abandon the analyzer, and put in its place the
piece of Iceland spar with which we have already illus-

trated double refraction. The two images of the carbon-points are now before you, produced, as you know,
by two beams vibrating at right angles to each other.
Introducing a plate of quartz between the polarizer
and the spar, the two images glow with complementary
colours. Employing the image of an aperture instead
of that of the carbon-points, we have two coloured circles. As the analyzer is caused to rotate, the colours
pass through various changes; but they are always
complementary to each other. When the one is red,
the other is green; when the one is yellow, the other
is blue. Here we have it in our power to demonstrate

Fig. 45.

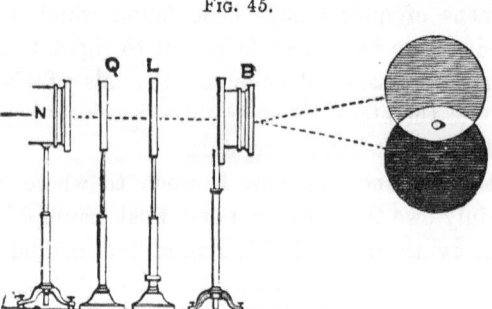

afresh a statement made in our first lecture, that,
although the mixture of blue and yellow pigments produces green, the mixture of blue and yellow lights
produces white. By enlarging our aperture, the two
images produced by the spar are caused to approach
each other, and finally to overlap. The one is now a
vivid yellow, the other a vivid blue, and you notice
that where the colours are superposed we have a pure
white. (See fig. 45, where N is the nozzle of the
lamp, Q the quartz plate, L a lens, and B the birefract-

ing spar. The two images overlap at O, and produce white by their mixture.)

This brings us to a point of our inquiries which, though rarely illustrated in lectures, is nevertheless so likely to affect profoundly the future course of scientific thought that I am unwilling to pass it over without reference. I refer to the experiment which Faraday, its discoverer, called the *magnetization of light*. The arrangement for this celebrated experiment is now before you. We have first our electric lamp, then a Nicol prism, to polarize the beam emergent from the lamp ; then an electro-magnet, then a second Nicol prism, and finally our screen. At the present moment the prisms are crossed, and the screen is dark. I place from pole to pole of the electro-magnet a cylinder of a peculiar kind of glass, first made by Faraday, and called Faraday's heavy glass. Through this glass the beam from the polarizer now passes, being intercepted by the Nicol in front. I now excite the magnet, and instantly light appears upon the screen. On examination, we find that, by the action of the magnet upon the ether contained within the heavy glass, the plane of vibration is caused to rotate, thus enabling the light to get through the analyzer.

The two classes into which quartz-crystals are divided have been already mentioned. In my hand I hold a compound plate, one half of it taken from a right-handed and the other from a left-handed crystal. Placing the plate in front of the polarizer, we turn one of the Nicols until the two halves of the plate show a common puce colour. This yields an exceedingly sensitive means of rendering visible the action of a magnet upon light. By turning either the polarizer or the

L

analyzer through the smallest angle, the uniformity of
the colour disappears, and the two halves of the quartz
show different colours. The magnet also produces this
effect. The puce-coloured circle is now before you on
the screen. (See fig. 46, where N is the nozzle of the

Fig. 46.

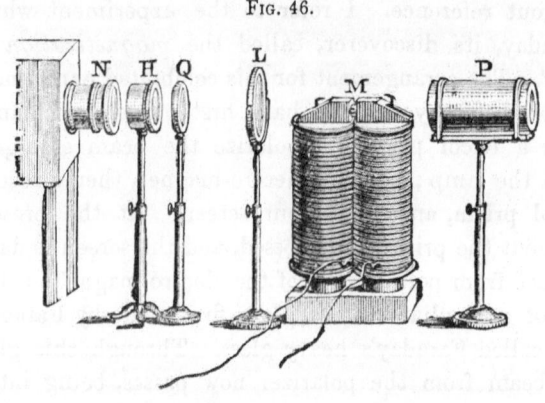

lamp, H the first Nicol, Q the biquartz plate, L a lens,
M the electro-magnet, with the heavy glass across
its poles, and P the second Nicol.) Exciting the magnet, one half of the image becomes suddenly red, the
other half green. Interrupting the current, the two
colours fade away, and the primitive puce is restored.
The action, moreover, depends upon the polarity of the
magnet, or, in other words, on the direction of the current which surrounds the magnet. Reversing the current, the red and green reappear, but they have changed
places. The red was formerly to the right, and the
green to the left; the green is now to the right, and
the red to the left. With the most exquisite ingenuity,
Faraday analyzed all those actions and stated their

laws. This experiment, however, long remained rather a scientific curiosity than a fruitful germ. That it would bear fruit of the highest importance, Faraday felt profoundly convinced, and recent researches are on the way to verify his conviction.

A few words more are necessary to complete our knowledge of the wonderful interaction between ponderable molecules and the ether interfused among them. Symmetry of molecular arrangement implies symmetry on the part of the ether; atomic dissymmetry, on the other hand, involves the dissymmetry of the ether, and, as a consequence, double refraction. In a certain class of crystals the structure is homogeneous, and such crystals produce no double refraction. In certain other crystals the molecules are ranged symmetrically round a certain line, and not around others. Along the former, therefore, the ray is undivided, while along all the others we have double refraction. Ice is a familiar example: its molecules are built with perfect symmetry around the perpendiculars to the planes of freezing, and a ray sent through ice in this direction is not doubly refracted; whereas, in all other directions, it is. Iceland spar is another example of the same kind: its molecules are built symmetrically round the line uniting the two blunt angles of the rhomb. In this direction a ray suffers no double refraction, in all others it does. This direction of no double refraction is called the *optic axis* of the crystal.

Hence, if a plate be cut from a crystal of Iceland spar perpendicular to the axis, all rays sent across this plate in the direction of the axis will produce but one image. But, the moment we deviate from the parallelism with the axis, double refraction sets in. If, therefore, a

beam that has been rendered *conical* by a converging
lens be sent through the spar so that the central ray of
the cone passes along the axis, this ray only will escape
double refraction.   Each of the others will be divided
into an ordinary and an extraordinary ray, the one
moving more slowly through the crystal than the
other ; the one, therefore, retarded with reference to
the other.   Here, then, we have the conditions for
interference, when the waves are reduced by the analyzer to a common plane.

Placing the plate of spar between the crossed Nicol's
prisms, and employing the conical beam, we have upon

Fig. 47.

the screen a beautiful system of iris rings surrounding the end of the optic axis, the circular
bands of colour being intersected by a black cross
(fig. 47). The arms of this cross are parallel to the two
directions of vibration in the polarizer and analyzer.
It is easy to see that those rays whose planes of vibration within the spar coincide with the plane of vibration
of *either* prism, cannot get through *both*. This complete interception produces the arms of the cross.
With monochromatic light the rings would be simply

bright and black—the bright rings occurring at those thicknesses of the spar which cause the rays to conspire; the black rings at those thicknesses which cause them to quench each other.   Turning the analyser 90°

FIG. 48.

round, we obtain the complementary phenomena.   The black cross gives place to a bright one, and every dark ring is supplanted also by a bright one (fig. 48.)   Here,

FIG. 49.

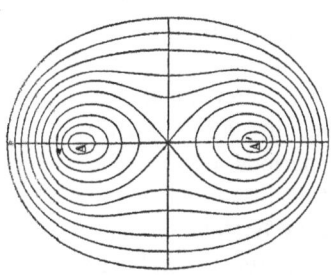

as elsewhere, the different lengths of the light-waves give rise to iris-colours when white light is employed.

Besides the *regular* crystals which produce double refraction in no direction, and the *uniaxal* crystals which produce it in all directions but one, Brewster discovered that in a large class of crystals there are *two* directions in which double refraction does not take place. These are called *biaxal* crystals. When plates of these crystals, suitably cut, are placed between the polarizer and analyzer, the axes (A A', fig. 49) are seen surrounded, not by circles, but by curves of another order and of a perfectly definite mathematical character. Each band, as proved experimentally by Herschel, forms a *lemniscata*; but the experimental proof was here, as in numberless other cases, preceded by the deduction which showed that, according to the undulatory theory the bands must possess this special character.

I have taken this somewhat wide range over polarization itself, and over the phenomena exhibited by crystals in polarized light, in order to give you some notion of the firmness and completeness of the theory which grasps them all. Starting from the single assumption of transverse undulations, we first of all determine the wave-lengths, and find all the phenomena of colour dependent on this element. The wave-lengths may be determined in many independent ways. Newton virtually determined them when he measured the periods of his Fits: the length of a fit, in fact, is that of a quarter of an undulation. The wave-lengths may be determined by diffraction at the edges of a slit (as in the Appendix); they may be deduced from the interference fringes produced by reflection; from the fringes produced by refraction; also by lines drawn with a diamond upon glass at measured distances asunder. And when the lengths determined by these independent methods are compared together, the strictest agreement is found to exist between them.

With the wave-lengths at our disposal, we follow the
ether into the most complicated cases of interaction
between it and ordinary matter, ' the theory is equal
to them all.   It makes not a single new physical
hypothesis ; but out of its original stock of principles
it educes the counterparts of all that observation shows.
It accounts for, explains, simplifies the most entangled
cases ; corrects known laws and facts ; predicts and dis-
closes unknown ones ; becomes the guide of its former
teacher Observation ; and, enlightened by mechanical
conceptions, acquires an insight which pierces through
shape and colour to force and cause.' [1]

But, while I have thus endeavoured to illustrate before you the power of the undulatory theory as a solver
of all the difficulties of optics, do I therefore wish you
to close your eyes to any evidence that may arise
against it ?   By no means.   You may urge, and justly
urge, that a hundred years ago another theory was held
by the most eminent men, and that, as the theory then
held had to yield, the undulatory theory may have to
yield also.   This seems reasonable ; but let us understand the precise value of the argument.   In similar
language a person in the time of Newton, or even in
our time, might reason thus : Hipparchus and Ptolemy,
and numbers of great men after them, believed that
the earth was the centre of the solar system.   But this
deep-set theoretic notion had to give way, and the
theory of gravitation may, in its turn, have to give
way also.   This is just as reasonable as the first argument.   Wherein consists the strength of the theory of
gravitation ?   Solely in its competence to account for
all the phenomena of the solar system.   Wherein consists the strength of the theory of undulation ?   Solely

[1] Whewell.

in its competence to disentangle and explain phenomena a hundred-fold more complex than those of the solar system. Accept if you will the scepticism of Mr. Mill[1] regarding the undulatory theory ; but if your scepticism be philosophical, it will wrap the theory of gravitation in the same or greater doubt.[2]

I am unwilling to quit these chromatic phenomena without referring to a source of colour which has often come before me of late in the blue of your skies at noon, and the deep crimson of your horizon after the set of sun. I will here summarise and extend what has been already said upon this subject. Proofs of the most cogent description could be adduced to show that the blue light of the firmament is reflected light. The light of the firmament comes to us across the direction of the solar rays, and even against the direction of the solar rays ; and this lateral and opposing rush of wave-motion can only be due to the rebound of the waves from the air itself, or from something suspended in the air. The solar light, moreover, is not reflected by the sky in the proportions which produce white. The sky is blue, which indicates an excess of the smaller waves. The blueness of the air has been given as a reason for the blueness of the sky ; but then the question arises, How, if the air be blue, can the light of sunrise and sunset, which travels through vast distances of air, be yellow, orange, or even red ? The passage of the white solar light through a blue medium could by no possibility

---

[1] Removed from us since these words were written.

[2] The only essay known to me on the Undulatory Theory, from the pen of an American writer, is an excellent one by President Barnard, published in the Smithsonian Report for 1862.

redden the light; the hypothesis of a blue air is therefore untenable. In fact the agent, whatever it be,
which sends us the light of the sky, exercises in so
doing a dichroitic action. The light reflected is blue,
the light transmitted is orange or red. A marked distinction is thus exhibited between reflection from the
sky and that from an ordinary cloud, which latter exercises no such dichroitic action.

The cloud, in fact, takes no note of size on the part
of the waves of ether, but reflects them all alike. Now
the cause of this may be that the cloud particles are so
large in comparison with the size of the waves of ether
as to scatter them all indifferently. A broad cliff reflects an Atlantic roller as easily as a ripple produced
by a sea-bird's wing ; and in the presence of large reflecting surfaces, the existing differences of magnitude
among the waves of ether may also disappear. But
supposing the reflecting particles, instead of being very
large, to be very small in comparison with the size of
the waves. Then, instead of the whole wave being
fronted and in great part thrown back, a small portion
only is shivered off by the obstacle. Suppose, then,
such minute foreign particles to be diffused in our atmosphere. Waves of all sizes impinge upon them,
and at every collision a portion of the impinging wave
is struck off. All the waves of the spectrum, from the
extreme red to the extreme violet, are thus acted upon;
but in what proportions will they be scattered ? Largeness is a thing of relation ; and the smaller the wave,
the greater is the relative size of any particle on which
the wave impinges, and the greater also the relative reflection.

A small pebble placed in the way of the ring-ripples

produced by heavy rain-drops on a tranquil pond will throw back a large fraction of each ripple incident upon it, while the fractional part of a larger wave thrown back by the same pebble might be infinitesimal. Now to preserve the solar light white, its constituent proportions must not be altered ; but in the scattering of the light by these very small particles we see that the proportions *are* altered. The smaller waves are in excess, and, as a consequence, in the scattered light blue will be the predominant colour. The other colours of the spectrum must, to some extent, be associated with the blue: they are not absent, but deficient. We ought, in fact, to have them all, but in diminishing proportions, from the violet to the red.

We have thus reasoned our way to the conclusion, that were particles, small in comparison to the size of the ether waves, sown in our atmosphere, the light scattered by those particles would be exactly such as we observe in our azure skies. And, indeed, when this light is analysed, all the colours of the spectrum are found in the proportions indicated by our conclusion.

By its successive collisions with the particles the white light is more and more robbed of its shorter waves; it therefore loses more and more of its due proportion of blue. The result may be anticipated. The transmitted light, where short distances are involved, will appear yellowish. But as the sun sinks towards the horizon the atmospheric distance increases, and consequently the number of the scattering particles. They weaken in succession the violet, the indigo, the blue, and even disturb the proportions of green. The transmitted light under such circumstances must pass

from yellow through orange to red. This also is
exactly what we find in nature. Thus, while the reflected light gives us at noon the deep azure of the
Alpine skies, the transmitted light gives us at sunset
the warm crimson of the Alpine snows.

But can small particles be really proved to act in the
manner indicated ? No doubt of it. Each one of you
can submit the question to an experimental test.
Water will not dissolve resin, but spirit will; and when
spirit which holds resin in solution is dropped into
water, the resin immediately separates in solid particles,
which render the water milky. The coarseness of this
precipitate depends on the quantity of the dissolved
resin. Professor Brücke has given us the proportions
which produce particles particularly suited to our present purpose. One gramme of clean mastic is dissolved
in eighty-seven grammes of absolute alcohol, and the
transparent solution is allowed to drop into a beaker
containing clear water briskly stirred. An exceedingly
fine precipitate is thus formed, which declares its
presence by its action upon light. Placing a dark-sur-
face behind the beaker, and permitting the light to fall
into it from the top or front, the medium is seen to be
of a very fair sky-blue. A trace of soap in water gives
a tint of blue. London, and I fear Liverpool, milk
makes an approximation to the same colour through
the operation of the same cause; and Helmholtz has
irreverently disclosed the fact that a blue eye is simply
a turbid medium.

But we have it in our power to imitate far more
closely the natural conditions of this problem. We can
generate in air artificial skies, and prove their perfect
identity with the natural one, as regards the exhibition

of a number of wholly unexpected phenomena. It has
been recently shown in a great number of instances that
waves of ether issuing from a strong source, such as the
sun or the electric light, are competent to shake asunder the atoms of gaseous molecules. The apparatus
used to illustrate this consists of a glass tube about a
yard in length, and from $2\frac{1}{2}$ to 3 inches internal diameter. The gas or vapour to be examined is introduced
into this tube, and upon it the condensed beam of the
electric lamp is permitted to act. The vapour is so
chosen that *one* at least of its products of decomposition,
as soon as it is formed, shall be *precipitated* to a kind
of cloud. By graduating the quantity of the vapour, this
precipitation may be rendered of any degree of fineness,
forming particles distinguishable by the naked eye, or
particles which are probably far beyond the reach of
our highest microscopic powers. I have no reason to
doubt that particles may be thus obtained whose
diameters constitute but a very small fraction of the
length of a wave of violet light.

Now, in all such cases when suitable vapours are
employed in a sufficiently attenuated state, no matter
what the vapour may be, the visible action commences
with the formation of a *blue cloud*. Let me guard myself at the outset against all misconception as to the use
of this term. The blue cloud here referred to is totally
invisible in ordinary daylight. To be seen, it requires
to be surrounded by darkness, *it only* being illuminated
by a powerful beam of light. This cloud differs in
many important particulars from the finest ordinary
clouds, and might justly have assigned to it an intermediate position between these clouds and true cloudless vapour.

It is possible to make the particles of this *actinic* cloud grow from an infinitesimal and altogether ultra-microscopic size to particles of sensible magnitude; and by means of these, in a certain stage of their growth, we produce a blue which rivals, if it does not transcend, that of the deepest and purest Italian sky. Introducing into our tube a quantity of mixed air and nitrite of butyl vapour sufficient to depress the mercurial column of an air-pump one-twentieth of an inch, and adding a quantity of air and hydrochloric acid sufficient to depress the mercury half an inch further, through this compound and highly attenuated atmosphere is sent the beam of the electric light. Gradually within the tube arises a splendid azure, which strengthens for a time, reaches a maximum of depth and purity, and then, as the particles grow larger, passes into whitish blue. This experiment is representative, and it illustrates a general principle. Various other colourless substances of the most diverse properties, optical and chemical, might be employed for this experiment. The *incipient cloud* in every case would exhibit this superb blue ; thus proving to demonstration that particles of infinitesimal size, without any colour of their own, and irrespective of those optical properties exhibited by the substance in a massive state, are competent to produce the blue colour of the sky.

But there is another subject connected with our firmament, of a more subtle and recondite character than even its colour. I mean that 'mysterious and beautiful phenomenon,' the polarization of the light of the sky. Looking at various points of the blue firmament through a Nicol's prism, and turning the prism round its axis, we soon notice variations of brightness. In certain positions of the prism, and from certain points of the firmament, the light appears to be wholly transmitted, while it is only necessary to turn the prism round its axis through an angle of ninety degrees to materially diminish the intensity of the light. Experiments of this kind prove that the blue light sent to us by the firmament is polarized, and on close scrutiny it is also found that the direction of most perfect polarization is perpendicular to the solar rays. Were the heavenly azure like the ordinary light of the sun, the turning of the prism would have no effect upon it; it would be transmitted equally during the entire rotation of the prism. The light of the sky is in great part quenched, because it is in great part polarized.

The same phenomenon is exhibited in perfection by our actinic clouds, the only condition necessary to its production being the smallness of the particles. In all cases, and with all substances, the cloud formed at the commencement, when the precipitated particles are sufficiently fine, is *blue*. In all cases, moreover, this fine blue cloud polarizes *perfectly* the beam which illuminates it, the direction of polarization enclosing an angle of 90° with the axis of the illuminating beam.

It is exceedingly interesting to observe both the perfection and the decay of this polarization. For ten or fifteen minutes after its first appearance the light from a vividly illuminated incipient cloud, looked at horizontally, is absolutely quenched by a Nicol's prism with its longer diagonal vertical. But as the sky-blue is gradually rendered impure by the introduction of particles of too large a size, in other words, as real

clouds begin to be formed, the polarization begins to
deteriorate, a portion of the light passing through the
prism in all its positions, as it does in the case of sky-
light.   It is worthy of note that for some time after
the cessation of perfect polarization the *residual* light
which passes, when the Nicol is in its position of
minimum transmission, is of a gorgeous blue, the
whiter light of the cloud being extinguished.   When
the cloud texture has become sufficiently coarse to ap-
proximate to that of ordinary clouds, the rotation of
the Nicol ceases to have any sensible effect on the
quantity of the light discharged at right angles to the
beam.

The perfection of the polarization in a direction
perpendicular to the illuminating beam may be also
illustrated by the following experiment with any suitable vapour.   A Nicol's prism large enough to embrace
the entire beam of the electric lamp was placed
between the lamp and the experimental tube.   Sending the beam polarized by the Nicol through the
tube, I placed myself in front of it, my eye being on a
level with its axis, my assistant occupying a similar
position behind the tube.   The short diagonal of the
large Nicol was in the first instance vertical, the plane
of vibration of the emergent beam being therefore also
vertical.   As the light continued to act, a superb blue
cloud visible to both my assistant and myself was slowly
formed.   But this cloud, so deep and rich when looked at
from the positions mentioned, *utterly disappeared when
looked at vertically downwards, or vertically upwards.*
Reflection from the cloud was not possible in these
directions.   When the large Nicol was slowly turned
round its axis, the eye of the observer being on the

level of the beam, and the line of vision perpendicular
to it, entire extinction of the light emitted horizontally occurred where the longer diagonal of the
large Nicol was vertical. But now a vivid blue cloud
was seen when looked at downwards or upwards. This
truly fine experiment, which I should certainly have
made without suggestion, was, as a matter of fact, first
definitely suggested by a remark addressed to me in
a letter by Prof. Stokes.

All the phenomena of colour and of polarization
observable in the case of skylight are manifested by
those actinic clouds; and they exhibit additional phenomena which it would be neither convenient to
pursue, nor perhaps possible to detect, in the actual
firmament. They enable us, for example, to follow
the polarization from its first appearance on the barely
visible blue to its final extinction in the coarser cloud.
These changes, as far as it is now necessary to refer to
them, may be thus summed up :—

1. The actinic cloud, as long as it continues blue,
discharges polarized light in all directions, but the
direction of maximum polarization, like that of sky-
light, is at right angles to the direction of the illuminating beam.

2. As long as the cloud remains distinctly blue the
light discharged from it at right angles to the illuminating beam is *perfectly* polarized. It may be utterly
quenched by a Nicol's prism, the cloud from which it
issues being caused to disappear. Any deviation from
the perpendicular enables a portion of the light to get
through the prism.

3. The direction of vibration of the polarized light
is at right angles to the illuminating beam. Hence a

plate of tourmaline, with its axis parallel to the beam, stops the light, and with the axis parallel to the beam transmits the light.

4. A plate of selenite placed between the Nicol and the actinic cloud shows the colours of polarized light; in fact, the cloud itself plays the part of a polarizing Nicol.

5. The particles of the blue cloud are immeasurably small, but they grow gradually in size, and at a certain period of their growth cease to discharge perfectly polarized light. For some time afterwards the light that reaches the eye through the Nicol is of a magnificent blue, far exceeding in depth and purity that of the purest sky; thus the waves that first *feel the influence of size* at both limits of the polarization are the shortest waves of the spectrum. These are the first to accept polarization, and they are the first to escape from it.

LECTURE V.

THE first question that we have to consider to-night
is this : Is the eye, as an organ of vision, commensurate
with the whole range of solar radiation—is it capable
of receiving visual impressions from all the rays emitted
by the sun ? The answer is negative. If we allowed
ourselves to accept for a moment that notion of gradual
growth, amelioration, and ascension, implied by the
term *evolution*, we might fairly conclude that there
are stores of visual impressions awaiting man far
greater than those of which he is now in possession.
For example, Ritter discovered in 1801 that beyond the
extreme violet of the spectrum there is a vast efflux of
rays which are totally useless as regards our present
powers of vision. These ultra-violet waves, however,

though incompetent to awaken the optic nerve, can so shake the molecules of certain compound substances on which they impinge as to effect their decomposition.

But though the blue, violet, and ultra-violet rays can act thus upon certain substances, the fact is hardly sufficient to entitle them to the name of ' chemical rays' as distinguished from the other constituents of the spectrum. As regards their action upon the salts of silver and many other substances— such, for example, as those concerned in the production of the actinic clouds referred to in our last lecture— they perhaps merit this title ; but in the case of the grandest example of the chemical action of light— namely, the decomposition of carbonic acid in the leaves of plants, with which my eminent friend Dr. Draper has so indissolubly associated his name—the yellow rays were found most active.

There are substances, however, on which the violet and ultra-violet waves exert a special decomposing power ; and, by permitting the invisible spectrum to fall upon surfaces prepared with such substances, we reveal both the existence and the extent of the ultra-violet spectrum.

The method of exhibiting the action of the ultra-violet rays by their chemical action has been long known ; indeed, Thomas Young photographed the ultra-violet rings of Newton. We have now to demonstrate their presence in another way. As a general rule, bodies transmit light or absorb it, but there is a third case in which the light falling upon the body is neither transmitted nor absorbed, but converted into light of another kind. Professor Stokes, the occupant of the chair of Newton in the University of Cambridge, has demonstrated this change of one kind of light into another, and has pushed his experiments so far as to render the invisible rays visible.

A long list of substances examined by Stokes when excited by the invisible ultra-violet waves have been proved to emit *light*. You know the rate of vibration corresponding to the extreme violet of the spectrum; you are aware that to produce the impression of this colour, the retina is struck 789 millions of millions of times in a second. At this point, the retina ceases to be useful as an organ of vision, for though struck by waves of more rapid recurrence, they are incompetent to awaken the sensation of light. But when such non-visual waves are caused to impinge upon the molecules of certain substances—on those of sulphate of quinine, for example—they compel those molecules, or their constituent atoms, to vibrate; and the peculiarity is, that the vibrations thus set up are *of slower period* than those of the exciting waves. By this lowering of the rate of vibration through the intermediation of the sulphate of quinine, the invisible rays are brought within the range of vision. We shall subsequently have abundant opportunity for learning that transparency to the visible by no means involves transparency to the invisible rays. Our bisulphide of carbon, for example, which, employed in prisms, is so eminently suitable for experiments on the visual rays, is by no means so suitable for these ultra-violet rays. Flint glass is better, and rock crystal is still better than flint glass. A glass prism, however, will suit our present purpose.

Casting by means of such a prism a spectrum, not upon the white surface of our screen, but upon a sheet

of paper which has been wetted with a saturated
solution of the sulphate of quinine, and afterwards dried,
an obvious extension of the spectrum is revealed. We
have, in the first instance, a portion of the violet
rendered whiter and more brilliant; but, besides this,
we have the gleaming of the colour where in the case
of unprepared paper nothing is seen. Other substances
produce a similar effect; and a substance recently discovered by President Morton, and to which he has given
the name of *Thallene*, produces a very striking elongation of the spectrum, the new light generated being
of peculiar brilliancy.

Fluor spar and some other substances when raised to
a temperature still under redness emit light. During the
ages which have elapsed since their formation, this capacity of shaking the ether into visual tremors appears to
have been enjoyed by these substances. Light has been
potential within them all this time ; and, as well explained by Draper, the heat, though not itself of visual
intensity, can unlock the molecules so as to enable
them to exert the power of vibration which they possess.
This deportment of fluor spar determined Stokes in his
choice of a name for his great discovery : he called this
rendering visible of the ultra-violet rays *Fluorescence*.

By means of a deeply-coloured violet glass, we cut
off almost the whole of the *light* of our electric beam ;
but this glass is peculiarly transparent to the violet and
ultra-violet rays. The violet beam now crosses a large
jar filled with water. Into it I pour a solution of
sulphate of quinine : opaque clouds, to all appearance, instantly tumble downwards. Fragments of
horse-chestnut bark thrown upon the water also send
down beautiful cloudlike striæ. But these are not

clouds : there is nothing precipitated here : the observed action is an action of *molecules,* not of *particles.*
The medium before you is not a turbid medium, for
when you look through it at a luminous surface it is
perfectly clear.

If we paint upon a piece of paper a flower or a
bouquet with the sulphate of quinine, and expose it to
the full beam, scarcely anything is seen. But on interposing the violet glass, the design instantly flashes forth
in strong contrast with the deep surrounding violet.
Here is a most beautiful example of such a design prepared for me by President Morton with his thallene :
placed in the violet light it exhibits a peculiarly
brilliant fluorescence. From the experiments of Dr.
Bence Jones, it would seem that there is some substance in the human body resembling the sulphate of
quinine, which causes all the tissues of the body to be
more or less fluorescent. The crystalline lens of the
eye exhibits the effect in a very striking manner.
When, for example, I plunge my eye into this violet
beam, I am conscious of a whitish-blue shimmer filling
the space before me. This is caused by fluorescent
light generated in the eye itself ; looked at from without, the crystalline lens at the same time gleams
vividly.

Long before its physical origin was understood this
fluorescent light attracted attention. Boyle, as Sir
Charles Wheatstone has been good enough to point
out to me, describes it with great fullness and exactness. ' We have sometimes,' he says, ' found in the
shops of our druggists a certain wood which is there
called *Lignum Nephriticum,* because the inhabitants
of the country where it grows are wont to use the

infusion of it, made in fair water, against the stone in the kidneys. This wood may afford us an experiment which, besides the singularity of it, may give no small assistance to an attentive considerer towards the detection of the nature of colours. Take *Lignum Nephriticum*, and with a knife cut it into thin slices; put about a handful of these slices into two or three or four pounds of the purest spring water. Decant this impregnated water into a glass phial; and if you hold it directly between the light and your eye, you shall see it wholly tinted with an almost golden colour. But if you hold this phial from the light, so that your eye be placed betwixt the window and the phial, the liquid will appear of a deep and lovely ceruleous colour.'

'These,' he continues, 'and other phenomena which I have observed in this delightful experiment, divers of my friends have looked upon, not without some wonder; and I remember an excellent oculist, finding by accident in a friend's chamber a phial full of this liquor, which I had given that friend, and having never heard anything of the experiment, nor having anybody near him who could tell him what this strange liquor might be, was a great while apprehensive, as he presently afterwards told me, that some strange new distemper was invading his eyes. And I confess that the unusualness of the phenomenon made me very solicitous to find out the cause of this experiment; and though I am far from pretending to have found it, yet my enquiries have, I suppose, enabled me to give such hints as may lead your greater sagacity to the discovery of the cause of this wonder.'[1]

[1] Boyle's Works, Birch's edition, vol. i. pp. 729 and 730.

Goethe in his 'Farbenlehre' thus describes the fluorescence of horse-chestnut bark:—'Let a strip of fresh horse-chestnut bark be taken and clipped into a glass of water; the most perfect sky-blue will be immediately produced.'[1] Sir John Herschel first noticed and described the fluorescence of the sulphate of quinine, and showed that the light proceeded from a thin stratum of the solution adjacent to the surface where the light enters it. He showed, moreover, that the incident beam, although not sensibly weakened in luminous power, lost the power of producing the blue fluorescent light in transmission through the solution of sulphate of quinine. Sir David Brewster also worked at the subject; but to Stokes we are indebted not only for its expansion, but for its full and final explanation.

But the waves from our incandescent carbon-points appeal to another sense than that of vision. They not only produce light, but heat, as a sensation. The magnified image of the carbon-points is now upon the screen; and with a suitable instrument the heating power of the rays which form that image might be demonstrated. In this case, however, the heat is spread over too large an area to be intense. Pushing out the lens and causing a movable screen to approach our lamp, the image is seen to become smaller and smaller; the rays at the same time becoming more concentrated, until finally they are able to pierce black paper with a burning ring. Rendering the beam parallel and receiving it upon a concave mirror, the rays are brought to a focus: paper placed at the focus is

[1] Werke, b. xxix. p. 24.

caused to smoke and burn. This may be done by our
ordinary camera and lens, and by a concave mirror of
very moderate power.

We will now adopt stronger measures with the radiation from the electric lamp. In this larger camera of
blackened tin is placed a lamp, in all particulars
similar to those already employed. But instead of
gathering up the rays from the carbon-points by a
condensing lens, we gather them up by a concave mirror
(mm', fig. 50), silvered in front and placed behind the
carbons (P). By this mirror we can cause the rays to issue
through the orifice in front of the camera, either parallel
or convergent. They are now parallel, and therefore to

Fig. 50.

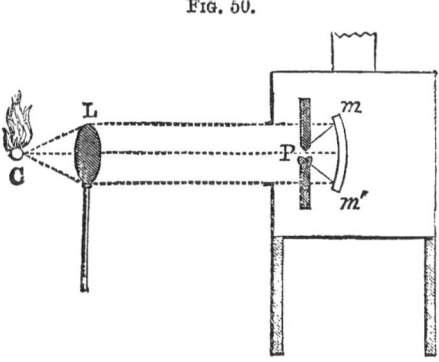

a certain extent diffused. We place a convex lens (L) in
the path of the beam ; the light is. converged to a
focus (C), and at that focus paper is not only pierced and
a burning ring formed, but it is instantly set ablaze.

Many metals may be burned up in the same way.
In our first lecture the combustibility of zinc was mentioned. Placing a strip of sheet-zinc at this focus, it

is instantly ignited and burns with its characteristic
purple flame.　And now I will substitute for our
glass lens (L) one of a more novel character.　In a
smooth iron mould this lens of pellucid ice has been
formed.　Placing it in the position occupied a moment
ago by the glass lens, I can see the beam brought to a
sharp focus.　At the focus I place a bit of black paper,
with a little gun-cotton folded up within it.　The paper
immediately ignites and the cotton explodes.　Strange,
is it not, that the beam should possess such heating
power after having passed though so cold a substance?
In his arctic expeditions Dr. Scoresby succeeded in
exploding gunpowder by the sun's rays converged by
large lenses of ice; here the effect is produced with a
small lens, and with a terrestrial source of heat.

In this experiment, you observe that, before the beam
reaches the ice-lens, it has passed through a glass cell
containing water.　The beam is thus sifted of constituents, which, if permitted to fall upon the lens,
would injure its surface, and blur the focus.　And this
leads me to say an anticipatory word regarding transparency.　In our first lecture we entered fully into the
production of colours by absorption, and we spoke repeatedly of the quenching of the rays of light.　Did
this mean that the light was altogether annihilated?
By no means.　It was simply so lowered in refrangibility as to escape the visual range.　*It was converted
into heat.*　Our red ribbon in the green of the spectrum
quenched the green, but if suitably examined its temperature would have been found raised.　Our green
ribbon in the red of the spectrum quenched the red,
but its temperature at the same time was augmented
to a degree exactly equivalent to the light extinguished.

Our black ribbon, when passed through the spectrum, was found competent to quench all its colours ; but at every stage of its progress an amount of heat was generated in the ribbon exactly equivalent to the light lost.   It is only when absorption takes [place that heat is thus produced; and heat is always a result of absorption.

Examine this water, then, in front of the lamp after the beam has passed through it : it is sensibly warm, and, if permitted to remain there long enough, it might be made to boil.   This is due to the absorption by the water of a portion of the electric beam.   But a certain portion passes through unabsorbed, and does not at all contribute to the heating of the water.   Now, ice is also in great part transparent to the latter portion, and therefore is but little melted by it ; hence, by employing this particular portion of the beam, we are able to keep our lens intact, and to produce by means of it a sharply-defined focus.[1]

Placed at that focus, white paper is not ignited, because it fails to absorb the rays emergent from the ice-lens.   At the same place, however, black paper instantly burns, because it absorbs the light which had passed through the ice-lens without absorption.   To these illustrations of heating power may be added the ignition of a diamond in oxygen, by the concentrated beam of the electric lamp.

---

[1] The comet of 1680, when nearest to the sun, was only a sixth of the sun's diameter from his surface.   Newton estimated its temperature here to be more than two thousand times that of molten iron.   It seems worth pointing out that the temperature of the comet could not be inferred from its nearness to the sun.   If its power of absorption were sufficiently low, the comet might carry into the sun's neighbourhood the temperature of stellar space.

In the path of the beam issuing from our lamp is
placed a cell with glass sides containing a solution of
alum. All the *light* of the beam passes through this
solution. This light is received on a powerfully converging mirror silvered in front, and brought to a focus
by the mirror. You can see the conical beam of reflected light tracking itself through the dust of the
room. A scrap of white paper placed at the focus
glows there with dazzling brightness, but it is not even
charred. On removing the alum cell, however, the
paper instantly inflames. There must, therefore, be
something in this beam besides its light. The *light* is
not absorbed by the white paper, and therefore does
not burn the paper; but there is something over and
above the light which *is* absorbed, and which provokes
combustion. What is this something?

In the year 1800 Sir William Herschel passed a
thermometer through the various colours of the solar
spectrum, and marked the rise of temperature corresponding to each colour. He found the heating effect
to augment from the violet to the red; he did not, however, stop at the red, but pushed his thermometer into
the dark space beyond it. Here he found the temperature actually higher than in any part of the visible
spectrum. By this important observation, he proved
that the sun emitted dark heat-rays which are entirely
unfit for the purposes of vision. The subject was subsequently taken up by Seebeck, Melloni, Müller, and
others, and within the last few years it has been found
capable of unexpected expansions and applications.
A method has been devised whereby the solar or electric
beam can be so *filtered* as to detach from it and preserve intact this invisible ultra-red emission, while the

visible and ultra-violet emissions are wholly intercepted. We are thus enabled to operate at will upon the purely ultra-red waves.

In the heating of solid bodies to incandescence this non-visual emission is the necessary basis of the visual. A platinum wire is stretched in front of the table, and through it an electric current flows. It is warmed by the current, and may be felt to be warm by the hand; it also emits waves of heat, but no light. Augmenting the strength of the current, the wire becomes hotter; it finally glows with a sober red light. At this point Dr. Draper many years ago began an interesting investigation. He employed a voltaic current to heat his platinum, and he studied by means of a prism the successive introduction of the colours of the spectrum. His first colour, as here, was red; then came orange, then yellow, then green, and lastly all the shades of blue. Thus as the temperature of the platinum was gradually augmented, the atoms were caused to vibrate more rapidly, shorter waves were thus produced, until finally waves were obtained corresponding to the entire spectrum. As each successive colour was introduced, the colours preceding it became more vivid. Now the vividness or intensity of light, like that of sound, depends not upon the length of the wave, but on the amplitude of the vibration. Hence, as the less refrangible colours grew more intense as the more refrangible ones were introduced, we are forced to conclude that side by side with the introduction of the shorter waves we had an augmentation of the amplitude of the longer ones.

These remarks apply not only to the visible emission examined by Dr. Draper, but to the visible emission which preceded the appearance of any light. In the emission from the white-hot platinum wire now before you the very waves exist with which we started, only their intensity has been increased a thousand-fold by the augmentation of temperature necessary to the production of this white light. Both effects are bound together: in an incandescent solid, or in a molten solid, you cannot have the shorter waves without this intensification of the longer ones. A sun is possible only on these conditions; hence Sir William Herschel's discovery of the invisible ultra-red solar emission.

The invisible heat, emitted both by dark bodies and by luminous ones, flies through space with the velocity of light, and is called *radiant heat*. Now, radiant heat may be made a subtle and powerful explorer of molecular condition, and of late years it has given a new significance to the act of chemical combination. Take, for example, the air we breathe. It is a mixture of oxygen and nitrogen; and it behaves towards radiant heat like a vacuum, being incompetent to absorb it in any sensible degree. But permit the same two gases to unite chemically; without any augmentation of the quantity of matter, without altering the gaseous condition, without interfering in any way with the *transparency* of the gas, the act of chemical union is accompanied by an enormous diminution of its *diathermancy*, or perviousness to radiant heat.

The researches which established this result also proved the elementary gases generally to be highly transparent to radiant heat. This, again, led to the proof of the diathermancy of elementary *liquids*, like bromine, and of *solutions* of the solid elements sulphur, phosphorus, and iodine. A spectrum is now before you, and you notice that this transparent bisulphide of carbon has no effect upon the colours. Dropping into the liquid a few flakes of iodine, you see the middle of the spectrum cut away. By augmenting the quantity of iodine, we invade the entire spectrum, and finally cut it off altogether. Now, the iodine which proves itself thus hostile to the light is perfectly transparent to the ultra-red emission with which we have now to deal. It, therefore, is to be our ray-filter.

Placing the alum-cell again in front of the electric lamp, we assure ourselves as before of the utter inability of the concentrated light to fire white paper. Introducing a cell containing the solution of iodine, the light is entirely cut off, and then on removing the alum-cell, the paper at the dark focus is instantly set on fire. Black paper is more absorbent than white for these ultra-red rays; and the consequence is, that with it the suddenness and vigour of the combustion are augmented. Zinc is burnt up at the same place, magnesium bursts into vivid combustion, while a sheet of platinized platinum placed at the focus is heated to whiteness.

Looked at through a prism, the white-hot platinum yields all the colours of the spectrum. Before impinging upon the platinum, the waves were of too slow recurrence to awaken vision; by the atoms of the platinum, these long and sluggish waves are broken up into shorter ones, being thus brought within the visual range. At the other end of the spectrum Stokes, by the interposition of suitable substances, *lowered* the refrangibility so as to render the non-visual rays visual, and to this change he gave the name of *Fluorescence*. Here, by the intervention of the platinum, the refrangibility is *raised*, so as to render the non-visual

visual, and to this change we give the name of *Calorescence.*

At the perfectly invisible focus where these effects are produced, the air may be as cold as ice. Air, as already stated, does not absorb the radiant heat, and is therefore not warmed by it. Nothing could more forcibly illustrate the isolation, if I may use the term, of the luminiferous ether from the air. The wave-motion of the one is heaped up, without sensible effect upon the other. I may add that, with suitable pre-cautions, the eye may be placed in a focus competent to heat platinum to vivid redness, without experiencing any damage, or the slightest sensation either of light or heat.

The important part played by these ultra-red rays in Nature may be thus illustrated : I remove the iodine filter, and concentrate the total beam upon a test-tube containing water. It immediately begins to sputter, and in a minute or two it *boils.* What boils it? Placing the alum solution in front of the lamp, the boiling instantly ceases. Now, the alum is pervious to all the luminous rays ; hence it cannot be these rays that caused the boiling. I now introduce the iodine, and remove the alum ; vigorous ebullition immediately recommences at the invisible focus. So that we here fix upon the invisible ultra-red rays the heating of the water.

We are thus enabled to understand the momentous part played by these rays in Nature. It is to them that we owe the warming and the consequent evapora-tion of the tropical ocean ; it is to them, therefore, that we owe our rains and snows. They are absorbed close to the surface of the ocean, and warm the superficial water, while the luminous rays plunge to great depths without producing any sensible effect. But we can proceed further than this. Here is a large flask containing a freezing mixture, which has so chilled the flask that the aqueous vapour of the air has been condensed and frozen upon it to a white fur. Introducing the alum-cell, and placing the coating of hoar-frost at the intensely luminous focus, not a spicula of the frost is melted. Introducing the iodine-cell, and removing the alum, a broad space of the frozen coating is instantly melted away. Hence we infer that the snow and ice which feed the Rhone, the Rhine, and other rivers which have glaciers for their sources, are released from their imprisonment upon the mountains by the invisible ultra-red rays of the sun.

The growth of science is organic. That which to-day is an *end* becomes to-morrow a *means* to a remoter end. Every new discovery in science is immediately made the basis of other discoveries, or of new methods of investigation. Thus about fifty years ago, Œrsted, of Copenhagen, discovered the deflection of a magnetic needle by an electric current; and about the same time Thomas Seebeck, of Berlin, discovered thermo-electricity. These great discoveries were soon afterwards turned to account by Nobili and Melloni in the construction of an instrument which has vastly augmented our knowledge of radiant heat. This instrument, which is called a *thermo-electric pile*, consists of thin bars of bismuth and antimony, soldered alternately together at their ends, but separated from each other elsewhere. From the ends of this 'pile' wires pass to a galvanometer, which consists of a coil of covered wire, within and above which are suspended two magnetic needles joined to a rigid system, and carefully defended from currents of air.

The action of the arrangement is this: the heat, falling on the pile, produces an electric current; the current, passing through the coil, deflects the needles, and the magnitude of the deflection may be made a measure of the heat. The upper needle moves over a graduated dial far too small to be seen. It is now, however, strongly illuminated. Above it is a lens which, if permitted, would form an image of the needle and dial upon the ceiling, where, however, it could not be conveniently seen. The beam is therefore received upon a looking-glass, placed at the proper angle, which throws the image upon a screen. In this way the motions of this small needle may be made visible to you all.

The delicacy of this apparatus is such that in a room filled, as this room now is, with an audience physically warm, it is exceedingly difficult to work with it. My assistant stands several feet off. I turn the pile towards him : the heat from his face, even at this distance, produces a deflection of 90°. I turn the instrument towards a distant wall, judged to be a little below the average temperature of the room. The needle descends and passes to the other side of zero, declaring by this negative deflection that the pile feels the chill of the wall. Possessed of this instrument, of our ray-filter, and of our large Nicol prisms, we are in a condition to investigate a subject of great philosophical interest, and which long engaged the attention of some of our foremost scientific workers—the substantial *identity of light and radiant heat.*

That they are identical in *all* respects cannot of

course be the case, for if they were so they would act
in the same manner upon all instruments, the *eye*
included.   The identity meant is such as subsists
between one colour and another, causing them to
behave alike as regards reflection, refraction, double
refraction, and polarization.   Let us here run rapidly
over their resemblances.   As regards reflection from
plane surfaces, we may employ a looking-glass to reflect
the light.   Marking any point in the track of the
reflected beam, and cutting off the light by the
iodine, on placing the pile at the marked point, the
needle immediately starts aside, showing that the heat
is reflected in the same direction.   This is true for
every position of the mirror.   Resuming, for example,
the experiments made with the apparatus employed in

Fig. 51.

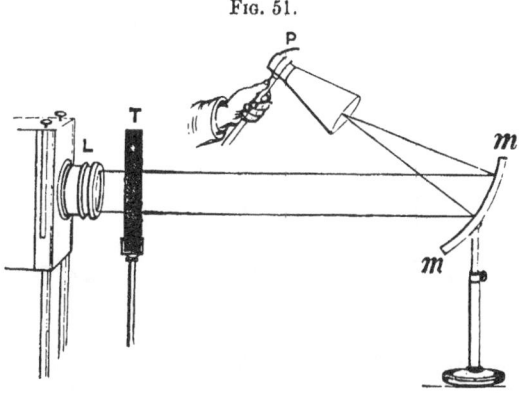

our first lecture (fig. 3); moving the index attached
to the mirror along the divisions of our graduated
arc (M O) and determining by the pile the positions
of the invisible reflected beam, we prove the angular
velocity of the beam to be twice that of the mirror.

As regards reflection from curved surfaces, the identity also holds good. Receiving the beam from our electric lamp on a concave mirror (m m, fig. 51), it is gathered up into a cone of reflected light; marking the apex of the cone by a pointer, and cutting off the light by the iodine solution (T), a moment's exposure of the pile (P) at the marked point produces a violent deflection of the needle.

The common and total reflection of a beam of radiant heat may be simultaneously demonstrated. From the nozzle of the lamp (L, fig. 52) a beam impinges upon

FIG. 52.

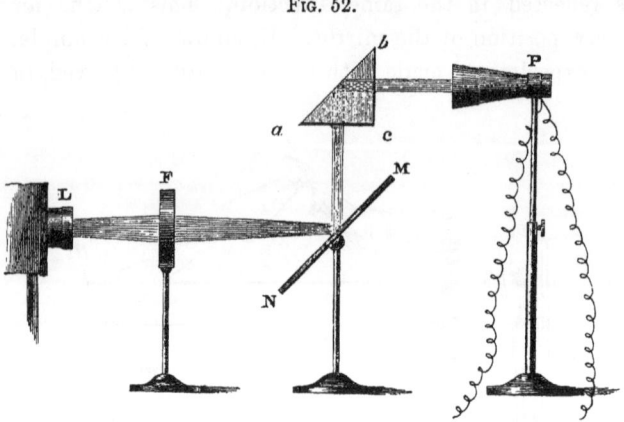

a plane mirror (M N), is reflected upwards, and enters a right-angled prism, of which a b c is the section. It meets the hypothenuse at an obliquity greater than the limiting angle,[1] and is therefore totally reflected. Quenching the light by the ray-filter at F, and placing the pile at P, the totally-reflected heat-

Defined in Lecture I.

beam is immediately felt by the pile, and declared by
the galvanometric deflection.

Perhaps no experiment more conclusively proves the
substantial identity of light and radiant heat than the
formation of invisible heat-images. Employing the
mirror already used to raise the beam to its highest
state of concentration, we obtain, as is well known, an
inverted image of the carbon points formed by the
light-rays at the focus. Cutting off the light by the
ray-filter, and placing at the focus a thin sheet of
platinized platinum, the invisible rays declare their

Fig. 53.

presence and distribution by stamping upon the platinum a white-hot image of the carbons. (See fig. 53.)

Whether radiant heat be capable of polarization or
not was for a long time a subject of discussion. Bérard
had announced affirmative results, but Powell and
Lloyd failed on trial to verify them. The doubts thus
thrown upon the question were removed by the experiments of Forbes, who first established the polarization
and 'depolarization' of heat. The subject was subsequently followed up by Melloni, an investigator of consummate ability, who sagaciously turned to account his

own discovery that the obscure rays of luminous sources were in part transmitted by black glass. Intercepting by a plate of this glass the light from his lamp oil flame, and operating upon the transmitted invisible heat, he obtained effects of polarization far exceeding in magnitude those which could be obtained with non-luminous sources. At present the possession of our more perfect ray-filter, and more powerful source of heat,

FIG. 54.

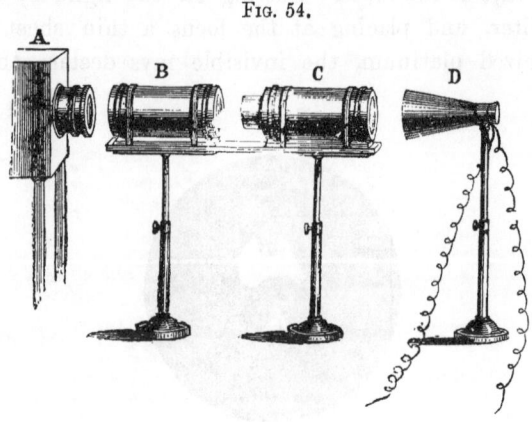

enables us to pursue this identity question to its utmost practical limits.

Mounting our two Nicols (B and C, fig. 54) in front of the electric lamp, with their principal sections crossed, no light reaches the screen. Placing our thermo-electric pile (D) behind the prisms, with its face turned towards the source, no deflection of the galvanometer is observed. Interposing between the lamp (A) and the first prism (B) our ray-filter, the light previously transmitted through the first Nicol is quenched; and now the slightest turning of either

Nicol opens a way for the transmission of the heat, a very small rotation sufficing to send the needle up to 90°. When the Nicol is turned back to its first position, the needle again sinks to zero, thus demonstrating in the plainest manner the polarization of the heat.

When the Nicols are crossed and the field is dark, you know, in the case of light, the effect of introducing a plate of mica between the polarizer and analyser. In two positions the mica exerts no sensible influence; in all others it does. A precisely analogous deportment is observed as regards radiant heat. Introducing our ray-filter, the thermo-pile, playing the part of an eye as regards the invisible radiation, receives no heat when the eye receives no light; but when the mica is so turned as to make its planes of vibration oblique to those of the polarizer and analyser, the heat immediately passes through. So strong does the action become, that the momentary plunging of the film of mica into the dark space between the Nicols suffices to send the needle up to 90°. This is the effect to which the term ' depolarization ' has been applied; the experiment really proving that with both light and heat we have the same resolution by the plate of mica, and recompounding by the analyser, of the ethereal vibrations.

Removing the mica and restoring the needle once more to 0°, I introduce between the Nicols a plate of quartz cut perpendicular to the axis; the immediate deflection of the needle declares the transmission of the heat, and when the transmitted beam is properly examined, it is found to be circularly polarized, exactly as a beam of light is polarized under the same conditions.

I will now abandon the Nicols, and send through

the piece of Iceland spar (B, fig. 55) already employed to illustrate the double refraction of light, our sifted beam

FIG. 55.

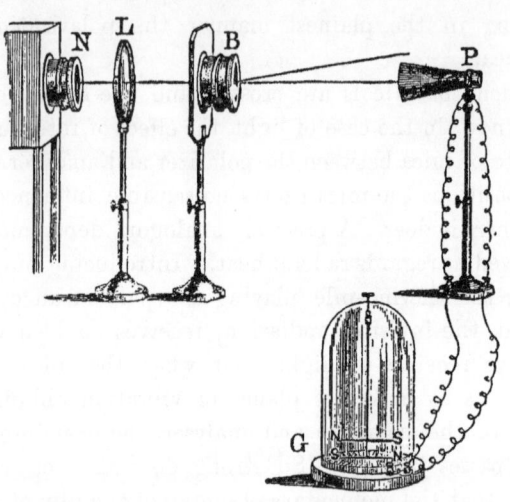

of invisible heat. To determine the positions of the two images, let us first operate upon the total beam. Marking the places of the light-images, we introduce (between N and L) our ray-filter (not in the figure) and quench the light. Causing the pile to approach one of the marked points, the needle remains unmoved until the point has been attained; here the pile at once detects the heat. Pushing the pile across the interval separating the two marks, the needle first falls to 0°, and then rises again to 90° in the second position.

I now turn the Iceland spar: the needle remains fixed: there is no alteration of the deflection. Passing the pile rapidly across to the other mark, the deflection is maintained. Once more I turn the spar,

but now the needle falls to 0°, rising, however, again
to 90° after a rotation of 360°. We know that in the
case of light the extraordinary beam rotates round the
ordinary one; and we have here been operating on the
extraordinary heat-beam, which, as regards double refraction, behaves exactly like a beam of light.

To render our series of comparisons complete, we must
demonstrate the magnetization of heat. But here a
slight modification of our arrangement will be necessary.
In repeating Faraday's experiment on the magnetization of light, we had, in the first instance, our Nicols
crossed and the field rendered dark, a flash of light appearing upon the screen when the magnet was excited.
Now the quantity of light transmitted in this case is
really very small, its effect being rendered striking
through contrast with the preceding darkness. When
we so place the Nicols that their principal sections enclose an angle of 45°, the excitement of the magnet
causes a far greater positive augmentation of the light,
though the augmentation is not so well *seen* through
lack of contrast, because here, at starting, the field is
illuminated.

In trying to magnetize our beam of heat, we will
adopt this arrangement. Here, however, at starting, a
considerable amount of heat falls upon one face of the
pile, which it is necessary to neutralize by permitting
rays from another source to fall upon the other face of
the pile. The needle is thus brought to zero. Cutting off the light by our ray-filter, and exciting the magnet, the needle is instantly deflected, proving that the
magnet has opened a door for the heat, exactly as in
Faraday's experiment it opened a door for the light.
Thus, in every case brought under our notice, the sub-

stantial identity of light and radiant heat has been demonstrated.

By the refined experiments of Knoblauch, who worked long and successfully at this question, the double refraction of heat by Iceland spar was first demonstrated ; but though he employed the luminous heat of the sun, the observed deflections were exceedingly small. So, likewise, those eminent investigators De la Povostaye and Desains succeeded in magnetizing a beam of heat; but though in their case also the luminous solar heat was employed, the deflection obtained did not amount to more than two or three degrees. With the arrangement here made use of deflections may be obtained, with purely invisible heat, equal to 150 of the lower degrees of the galvanometer.

We have finally to determine the position and magnitude of the invisible radiation which produces these results. For this purpose we employ a particular form of the thermo-electric pile. Its face is a rectangle, which by movable side-pieces can be rendered as narrow as desirable. Throwing a small and concentrated spectrum upon a screen, by means of an endless screw we move this rectangular pile through the entire spectrum, and determine in succession the thermal power of all its colours.

When this instrument is brought to the violet end of the spectrum, the heat is found to be almost insensible. As the pile gradually moves from the violet towards the red, it encounters a gradually augmenting heat. The red itself possesses the highest heating power of all the colours of the spectrum. Pushing the pile into the dark space beyond the red, the heat rises suddenly in intensity, and at some distance beyond the red it attains a maximum. From this point the heat falls somewhat more rapidly than it rose, and afterwards gradually fades away.

Drawing a horizontal line to represent the length of the spectrum, and erecting along it, at various points, perpendiculars proportional in length to the heat existing at those points, we obtain a curve which exhibits the distribution of heat in our spectrum. It is represented in the adjacent figure. Beginning at the blue, the curve rises, at first very gradually; towards the red it rises more rapidly, the line C D (fig. 56, next page) representing the strength of the extreme red radiation. Beyond the red it shoots upwards in a steep and massive peak to B, whence it falls, rapidly for a time, and afterwards gradually fades from the perception of the pile. This figure is the result of more than twelve careful series of measurements, for each of which the curve was constructed. On superposing all these curves, a satisfactory agreement was found to exist between them. So that it may safely be concluded that the areas of the dark and white spaces respectively represent the relative energies of the visible and invisible radiation. The one is 7·7 times the other.

But in verification, as already stated, consists the strength of science. Determining in the first place the total emission from the electric lamp; then by means of the iodine filter determining the ultra-red emission; the difference between both gives the luminous emission. In this way, it is found that the energy of the invisible emission is eight times that of the visible. No two methods could be more opposed to each other, and hardly any two results could better harmonize. I

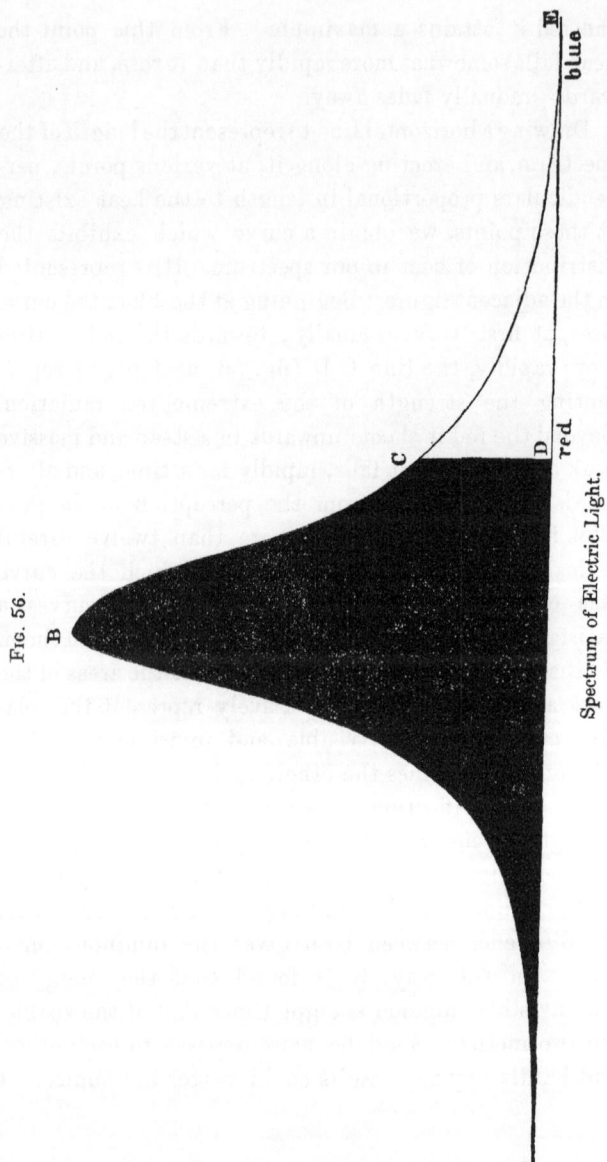

Fig. 56.

Spectrum of Electric Light.

think, therefore, you may rely upon the accuracy of the distribution of heat here assigned to the prismatic spectrum of the electric light. There is nothing vague in the mode of investigation, nor doubtful in its conclusions.

LECTURE VI.

PRINCIPLES OF SPECTRUM ANALYSIS—PRISMATIC ANALYSIS OF THE LIGHT
OF INCANDESCENT VAPOURS—DISCONTINUOUS SPECTRA—SPECTRUM
BANDS PROVED BY BUNSEN AND KIRCHHOFF TO BE CHARACTERISTIC OF
THE VAPOUR—DISCOVERY OF RUBIDIUM, CÆSIUM, AND THALLIUM—
RELATION OF EMISSION TO ABSORPTION—THE LINES OF FRAUNHOFER
—THEIR EXPLANATION BY KIRCHHOFF—SOLAR CHEMISTRY INVOLVED
IN THIS EXPLANATION — FOUCAULT'S EXPERIMENT — PRINCIPLES OF
ABSORPTION—ANALOGY OF SOUND AND LIGHT—EXPERIMENTAL DE-
MONSTRATION OF THIS ANALOGY — RECENT APPLICATIONS OF THE
SPECTROSCOPE—SUMMARY AND CONCLUSION.

WE have employed as our source of light in these
lectures the ends of two rods of coke rendered incandescent by electricity. Coke is particularly suitable for
this purpose, because it can bear intense heat without
fusion or vaporization. It is also black, which helps the
light; for, other circumstances being equal, as shown
experimentally by Professor Balfour Stewart, the blacker
the body the brighter will be its light when incandescent. Still, refractory as carbon is, if we closely examined our voltaic arc, or stream of light between the
carbon-points, we should find there incandescent carbon-
vapour. And if we could detach the light of this vapour
from the more dazzling light of the solid points, we
should find its spectrum not only less brilliant, but of a
totally different character from the spectra that we have
already seen. Instead of being an unbroken succession
of colours from red to violet, the carbon-vapour would

yield a few bands of colour with spaces of darkness between them.

What is true of the carbon is true in a still more striking degree of the metals, the most refractory of which can be fused, boiled, and reduced to vapour by the electric current. From the incandescent vapour the light, as a general rule, flashes in groups of rays of definite degrees of refrangibility, spaces existing between group and group, which are unfilled by rays of any kind. But the contemplation of the facts will render this subject more intelligible than words can make it. Within the camera is now placed a cylinder of carbon hollowed out at the top to receive a bit of metal; in the hollow is placed a fragment of the metal thallium. Down upon this we bring the upper carbon point, and then separate the one from the other. A stream of incandescent thallium vapour passes between them, the magnified image of which is now seen upon the screen. It is of a beautiful green colour. What is the meaning of that green? We answer the question by subjecting the light to prismatic analysis. Here you have its spectrum, consisting of a single refracted band. Light of one degree of refrangibility, and that corresponding to green, is emitted by the thallium vapour.

We will now remove the thallium and put a bit of silver in its place. The arc of silver is not to be distinguished from that of thallium; it is not only green, like the thallium vapour, but the same shade of green. Are they then alike? Prismatic analysis enables us to answer the question. It is perfectly impossible to confound the spectrum of incandescent silver vapour with that of thallium. Here are two green bands instead of one.

Adding to the silver in our camera a bit of thallium, we shall obtain the light of both metals, and after waiting a little we see that the green of the thallium lies midway between the two greens of the silver. Hence this similarity of colour.

But you observe another interesting fact. The thallium band is at first far brighter than the silver bands. Indeed, the latter have wonderfully degenerated since the bit of thallium was put in, and for a reason worth knowing. It is the *resistance* offered to the passage of the electric current from carbon to carbon that calls forth the power of the current to produce heat. If the resistance were materially lessened, the heat would be materially lessened; and if all resistance were abolished, there would be no heat at all. Now, thallium is a much more fusible and vaporizable metal than silver; and its vapour facilitates the passage of the current to such a degree as to render it almost incompetent to vaporize the more refractory silver. But the thallium is gradually consumed; its vapour diminishes, the resistance rises, until finally you see the two silver bands as brilliant as they were at first.[1]

We have in these bands a perfectly unalterable characteristic of these two metals. You never get other bands than these two green ones from the silver, never other than the single green band from the thallium, never other than the three green bands from the mixture of both metals. Every known metal has its own particular bands, and in no known case are the bands of two different metals alike in refrangibility.

---

[1] This circumstance ought not to be lost sight of in the examination of compound spectra.

It follows, therefore, that these spectra may be made a sure test for the presence or absence of any particular metal. If we pass from the metals to their alloys, we find no confusion. Copper gives green bands; zinc gives blue and red bands; brass, an alloy of copper and zinc, gives the bands of both metals, perfectly unaltered in position or character.

But we are not confined to the metals themselves; the *salts* of these metals yield the bands of the metals. Chemical union is ruptured by a sufficiently high heat; the vapour of the metal is set free and yields its characteristic bands. The chlorides of the metals are particularly suitable for experiments of this character. Common salt, for example, is a compound of chlorine and sodium; in the electric lamp it yields the spectrum of the metal sodium. The chlorides of copper, lithium, and strontium yield in like manner the bands of these metals.

When, therefore, Bunsen and Kirchhoff, the celebrated founders of *spectrum analysis*, after having established by an exhaustive examination the spectra of all known substances, discovered a spectrum containing bands different from any known bands, they immediately inferred the existence of a new metal. They were operating at the time upon a residue obtained by evaporating one of the mineral waters of Germany. In that water they knew the unknown metal was concealed, but vast quantities of it had to be evaporated before a residue could be obtained sufficiently large to enable ordinary chemistry to grapple with the metal. They, however, hunted it down, and it now stands among chemical substances as the metal *Rubidium*. They subsequently discovered a second metal, which they called *Cæsium*. Thus, having first

o

placed spectrum analysis on a sure foundation, they demonstrated its capacity as an agent of discovery. Soon afterwards Mr. Crookes, pursuing the same method, discovered the bright green band of thallium, and obtained the salts of the metal which yielded it. The metal itself was first isolated in ingots by M. Lamy, a French chemist.

All this relates to chemical discovery upon earth, where the materials are in our own hands. But it was soon shown how spectrum analysis might be applied to the investigation of the sun and stars; and this result was reached through the solution of a problem which had been long an enigma to natural philosophers. The scope and conquest of this problem we must now endeavour to comprehend. A spectrum is *pure* in which the colours do not overlap each other. We purify the spectrum by making our slit narrow and by augmenting the number of our prisms. When a pure spectrum of the sun has been obtained in this way it is found to be furrowed by innumerable dark lines. Four of them were first seen by Dr. Wollaston, but they were afterwards multiplied and measured by Fraunhofer with such masterly skill that they are now universally known as Fraunhofer's lines. To give an explanation of these lines was, as I have said, a problem which long challenged the attention of philosophers, and to Kirchhoff, Professor of Physics in the University of Heidelberg, belongs the honour of having first conquered this problem.

(The positions of the principal lines, lettered according to Fraunhofer, are shown in the annexed sketch (fig. 57) of the solar spectrum. A is supposed to stand near the extreme red, and J near the extreme violet.)

The brief memoir of two pages on which this immortal discovery is recorded was communicated to
the Berlin Academy on October 27, 1859.   FIG. 57.
Fraunhofer had remarked in the spectrum
of a candle flame two bright lines which
coincide accurately as to position with the
double dark line D of the solar spectrum.
These bright lines are produced with particular intensity by the yellow flame derived
from a mixture of salt and alcohol. They
are in fact the lines of sodium vapour.
Kirchhoff produced a spectrum by permitting the sunlight to enter his telescope by
a slit and prism, and in front of the slit
he placed the yellow sodium flame.   As
long as the spectrum remained feeble, there
always appeared two bright lines, derived
from the flame, in the place of the two dark
lines D of the spectrum.   In this case such
absorption as the flame exerted upon the sunlight was more than atoned for by the radiation from the flame.   When, however, the
solar spectrum was rendered sufficiently intense, the bright bands entirely vanished
and the two dark Fraunhofer lines appeared
with much greater sharpness and distinctness than when the flame was not employed.

This result, be it noted, was not due to
any real quenching of the bright lines of the
flame, but to the augmentation of the intensity of the adjacent spectrum. The experiment proved to demonstration that when the white light
sent through the flame was sufficiently intense, the

quantity which the flame absorbed was far in excess of the quantity which it radiated.

Here then is a result of the utmost significance. Kirchhoff immediately inferred from it that the salt flame which could intensify so remarkably the dark lines of Fraunhofer ought also to be able to *produce* them. The spectrum of the Drummond light is known to show the two bright lines of sodium, which, however, gradually disappear as the modicum of sodium contained, as an impurity, in the incandescent lime is exhausted. Kirchhoff formed a spectrum of the lime-light, and after the two bright lines had vanished, he placed his salt flame in front of the slit. The two dark lines D immediately started forth. Thus in the continuous spectrum of the lime-light he evoked, artificially, the lines D of Fraunhofer.

Kirchhoff knew that this was an action not peculiar to the sodium flame, and he immediately extended his result to all coloured flames which yield sharply defined bands in their spectra. White light, with all its constituents complete, sent through such flames, would, he inferred, have those precise constituents absorbed the refrangibilities of which are the same as those of the bright bands ; so that after passing through such flames, the white light, if sufficiently intense, would have its spectrum furrowed by bands of darkness. On the occasion here referred to, Kirchhoff also succeeded in reversing a bright band of lithium.

The long-standing difficulty of Fraunhofer's lines fell to pieces in the presence of facts and reflections like these, which also carried with them an immeasurable extension of the chemist's power. Kirchhoff saw that from the lines in their spectra, in so far as they are shown by terrestrial substances, the presence or absence of these substances in the sun and fixed stars might be immediately inferred. Thus the dark lines D in the solar spectrum proved the existence of sodium vapour in the solar atmosphere ; while the bright lines discovered by Brewster in a nitre flame, which had been proved to coincide exactly with certain dark lines between A and B in the solar spectrum, proved the existence of potassium in the sun.

All subsequent research verified the accuracy of these first daring conclusions. In his second paper, communicated to the Berlin Academy before the close of 1859, Kirchhoff proved the existence of iron in the sun. The bright lines of the spectrum of iron vapour are exceedingly numerous, and 65 of them were subsequently proved by Kirchhoff to be absolutely identical in position with 65 dark Fraunhofer's lines. Ångstrom and Thalén pushed the coincidences to 450 for iron, while according to the same excellent investigators the following numbers express the coincidences in the case of the respective metals to which they are attached :—

Calcium .	.	. 75	Nickel	.	.	. 33
Barium .	.	. 11	Cobalt	.	.	. 19
Magnesium	.	. 4	Hydrogen	.	.	4
Manganese	.	. 57	Aluminium	.	.	2
Titanium .	.	. 118	Zinc	.	.	. 2
Chromium	.	. 18	Copper	.	.	. 7

The probability is overwhelming that all these substances exist in the atmosphere of the sun.

Kirchhoff's discovery profoundly modified the conceptions previously entertained regarding the constitution of the sun, and they led him to a view of that constitution which, though it may be modified in detail, will, I believe, remain substantially valid to the end of

time. The sun consists of a nucleus which is surrounded by a flaming atmosphere of lower temperature. That nucleus may, in part, be *clouds*, under-
lying true vapour. The light of the nucleus would
give us a continuous spectrum, as our carbon points
did; but having to pass through the photosphere,
as our beam through the sodium flame, those rays
of the nucleus which the photosphere can itself emit
are absorbed, and shaded lines, corresponding to
the particular rays absorbed, occur in the spectrum.
Abolish the solar nucleus, and we should have a spectrum showing a bright line in the place of every dark
line of Fraunhofer, just as in the case of Kirchhoff's
second experiment we should have the bright sodium
lines if the lime light were withdrawn. These lines
of Fraunhofer are therefore not absolutely dark, but
dark by an amount corresponding to the difference
between the light intercepted and the light emitted by
the photosphere.

Almost every great scientific discovery is approached
contemporaneously by many minds, the fact that one
mind usually confers upon it the distinctness of demonstration being an illustration not of genius isolated, but
of genius in advance. Thus Foucault, in 1849, came
to the verge of Kirchhoff's discovery. By converging
an image of the sun upon a voltaic arc, and thus obtaining the spectra of both sun and arc superposed,
he found that the two bright lines which, owing to the
presence of a little sodium impurity in the carbons, or
in the air, are seen in the spectrum of the arc, coincide
with the dark lines D of the solar spectrum. The lines
D he found to be considerably strengthened by the
passage of the solar light through the voltaic arc.

Instead of the image of the sun, he then projected upon the arc the image of one of the solid incandescent carbon points, which of itself would give a continuous spectrum, and he found that the lines D were thus *generated* in that spectrum. Foucault's conclusion from this admirable experiment was 'that the arc is a medium which emits the rays D on its own account, and at the same time absorbs them when they come from another quarter.' Here he stopped. He did not extend his observations beyond the voltaic arc; he did not offer any explanation of the lines of Fraunhofer; he did not arrive at any conception of solar chemistry, or of the constitution of the sun. His beautiful experiment remained a germ without fruit, until the discernment, ten years subsequently, of the whole class of phenomena to which it belongs enabled Kirchhoff to solve these great problems.

Soon after the publication of Kirchhoff's discovery Professor Stokes, who, ten years prior to the discovery, had nearly anticipated it, borrowed an illustration from sound to show the reciprocity of radiation and absorption. A stretched string responds to aërial vibrations which synchronize with its own. A great number of such strings stretched in space would roughly represent a medium; and if the note common to them all were sounded at a distance they would absorb the vibrations corresponding to that note. That is to say, they would absorb the vibrations which they can emit.

When a violin-bow is drawn across this tuning-fork, the room is immediately filled with a musical sound; this may be regarded as the *radiation* or *emission* of sound from the fork. A few days ago, on sounding this fork, I noticed that when its vibrations were quenched, the sound seemed to be continued, though more feebly. It appeared, moreover, to come from under a distant table, where stood a number of tuning-forks of different sizes and rates of vibration. One of these, and one only, had been started by the sounding fork, and it was one whose rate of vibration was the same as that of the fork which started it. This is an instance of the *absorption* of the sound of one fork by another. Placing two unisonant forks near each other, sweeping the bow over one of them, and then quenching the agitated fork, the other continues to sound; this other can re-excite the former, and several tranfers of sound between the two forks can be thus effected. Placing a cent-piece on each prong of one of the forks, we destroy its perfect synchronism with the other, and no communication of sound from the one to the other is then possible.

I have now to bring before you, on a suitable scale, the demonstration that we can do with *light* what has been here done with sound. For several days in 1861 I endeavoured to accomplish this with only partial success. In iron dishes a mixture of dilute alcohol and salt was placed and warmed so as to promote vaporization. The vapour was ignited, and through the yellow flame thus produced the beam from the electric lamp was sent; but a faint darkening only of the yellow band of a projected spectrum could be obtained. A trough was then made which, when fed with the salt and alcohol, yielded a flame ten feet thick; but the result of sending the light through this depth of flame was still unsatisfactory. Remembering that the direct combustion of sodium in a Bunsen's flame, produced a yellow far more intense than the salt flame, and inferring

that the intensity of the colour indicated the copious-
ness of the incandescent vapour, I sent through the
flame from metallic sodium the beam of the electric
lamp. The success was complete ; and this experiment
I wish now to repeat in your presence.[1]

Firstly then you notice, when a fragment of sodium
is placed in a tin spoon and introduced into a lightless
Bunsen's flame, an intensely yellow light is produced,
which corresponds in refrangibility with the yellow
band of the spectrum.   Like our tuning-fork, it emits
waves of a special period.   When the white light from
the electric lamp is sent through that flame, you will
have ocular proof that the yellow flame intercepts the
yellow of the spectrum ; in other words, that it absorbs
waves of the same period as its own, thus producing
to all intents and purposes a dark Fraunhofer's band
in the place of the yellow.

In front of the slit (at L, fig. 58) through which the
beam issues is placed a Bunsen's burner (b) protected by
a chimney (C). This beam, after passing through a lens,
traverses the prism (P) (in the real experiment there
was a pair of them), is there decomposed, and forms a
vivid continuous spectrum (S S) upon the screen.   Introducing a tin spoon with its pellet of sodium into the
Bunsen's flame, the metal first fuses, colours the flame
intensely yellow, and at length bursts into violent
combustion.   At the same moment the spectrum is
furrowed by a dark band (D).   Introducing and withdrawing the sodium flame in rapid succession, the

---

[1] The dark band produced when the sodium is placed within the
lamp was observed on the same occasion. Then was also observed for
the first time the magnificent blue band of lithium which the Bunsen's
flame fails to bring out.

sudden appearance and disappearance of the band of darkness is shown in a most striking manner. In contrast with the adjacent brightness this band appears absolutely black, so vigorous is the absorption. The blackness, however, is but relative, for upon the dark space falls a portion of the light of the sodium flame.

FIG. 58.

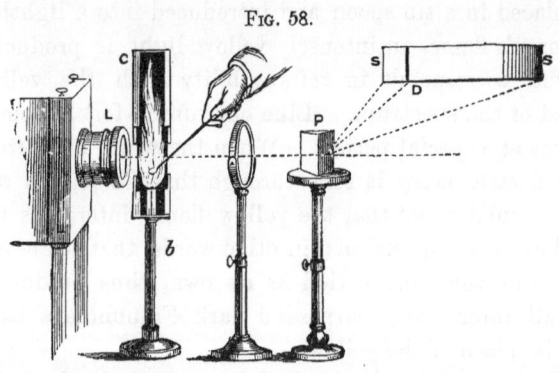

I have already referred to the experiment of Foucault; but other workers also had been engaged on the borders of this subject before it was taken up by Bunsen and Kirchhoff. With a few modifications here introduced, I have already spoken in this wise of the precursors of the discovery of spectrum analysis and solar chemistry :—'Mr. Talbot had observed the bright lines in the spectra of coloured flames, and both he and Sir John Herschel pointed out the possibility of making prismatic analysis a chemical test of exceeding delicacy, though not, it would appear, of entire certainty. More than a quarter of a century ago Dr. Miller gave drawings and descriptions of the spectra of various coloured flames. Wheatstone, with his accustomed acuteness, analyzed the light of the electric spark, and proved that the metals between which the spark passed determined the bright bands in the spectrum of the spark. In an investigation described by Kirchhoff as 'classical,' Swan had shown that $\frac{1}{2,500,000}$ of a grain of sodium in a Bunsen's flame could be detected by its spectrum. He also proved the constancy of the bright lines in the spectra of hydro-carbon flames. Masson published a prize essay on the bands of the induction spark; while Van der Willigen, and more recently Plücker, have also given us beautiful drawings of spectra obtained from the same source.

'But none of these distinguished men betrayed the least knowledge of the connection between the bright bands of the metals and the dark lines of the solar spectrum, nor could spectrum analysis be said to be placed upon anything like a safe foundation prior to the researches of Bunsen and Kirchhoff. The man who, in a published paper, came nearest to the philosophy of the subject was Ångström. In that paper translated by myself, and published in the "Philosophical Magazine" for 1855, he indicates that the rays which a body absorbs are precisely those which it can emit when rendered luminous. In another place, he speaks of one of his spectra giving the general impression of the *reversal* of the solar spectrum. But his memoir, philosophical as it is, is distinctly marked by the uncertainty of his time. Foucault, Thomson, and Balfour Stewart have all been near the discovery, while, as already stated, it was almost hit by the acute but unpublished conjecture of Stokes.'

Mentally, as well as physically, every year of the world's age is the outgrowth and offspring of all preceding

years. Science proves itself to be a genuine product
of Nature by growing according to this law. We have
no solution of continuity here. All great discoveries
are duly prepared for in two ways: first, by other discoveries which form their prelude ; and, secondly, by the
sharpening of the enquiring intellect. Thus Ptolemy
grew out of Hipparchus, Copernicus out of both, Kepler
out of all three, and Newton out of all the four. Newton did not rise suddenly from the sea-level of the
intellect to his amazing elevation. At the time that
he appeared, the table-land of knowledge was already
high. He juts, it is true, above the table-land, as a
massive peak ; still he is supported by it, and a great
part of his absolute height is the height of humanity
in his time. It is thus with the discoveries of Kirchhoff. Much had been previously accomplished ; this
he mastered, and then by the force of individual genius
went beyond it. He replaced uncertainty by certainty,
vagueness by definiteness, confusion by order ; and I
do not think that Newton has a surer claim to the
discoveries that have made his name immortal, than
Kirchhoff has to the credit of gathering up the frag-
mentary knowledge of his time, of vastly extending it,
and of infusing into it the life of great principles.

With one additional point we will wind up our
illustrations of the principles of solar chemistry.
Owing to the scattering of light by matter floating
mechanically in the earth's atmosphere, the sun is
seen not sharply defined, but surrounded by a luminous glare. Now, a loud noise will drown a whisper,
an intense light will quench a feeble one, and so this
circumsolar glare prevents us from seeing many striking
appearances round the border of the sun. The glare is

abolished in total eclipses, when the moon comes between the earth and the sun, and there are then seen a series of rose-coloured protuberances stretching sometimes tens of thousands of miles beyond the dark edge of the moon. They are described by Vassenius in the 'Philosophical Transactions' for 1733; and were probably observed even earlier than this. In 1842 they attracted great attention, and were then compared to Alpine snow-peaks reddened by the evening sun. That these prominences are flaming gas, and principally hydrogen gas, was proved by Mr. Janssen during an eclipse observed in India, on the 18th of August 1868.

But the prominences may be rendered visible in full sunshine; and for a reason easily understood. You have seen in these lectures a single prism employed to produce a spectrum, and you have seen a pair of prisms employed. In the latter case, the dispersed white light, being diffused over about twice the area, had all its colours proportionately diluted. You have also seen one prism and a pair of prisms employed to produce the bands of incandescent vapours; but here the light of each band, being absolutely monochromatic, was incapable of further dispersion by the second prism, and could not therefore be weakened by such dispersion.

Apply these considerations to the circumsolar region. The glare of white light round the sun can be dispersed and weakened to any extent by augmenting the number of prisms, while a monochromatic light, mixed with this glare, and masked by it, would retain its intensity unenfeebled by dispersion. Upon this consideration has been founded a method of observation, applied independently by M. Janssen in India and by Mr. Lockyer

in England, by which the monochromatic bands of the
prominences are caused to obtain the mastery, and to
appear in broad daylight.

It would lead us far beyond the object of these lectures
to dwell upon the numerous interesting and important
results obtained by Secchi, Respighi, Young, and the
other distinguished men who have worked at the
chemistry of the sun and its appendages. Nor can
I do more at present than make a passing reference
to the excellent labours of Dr. Huggins in connexion
with the fixed stars, nebulæ, and comets. They, more
than any others, illustrate the literal truth of the
statement, that the establishment of spectrum analysis
and the explanation of Fraunhofer's lines, carried with
them an immeasurable extension of the chemist's
range. But my object here is to make principles plain,
rather than to follow out the details of their illustration.
This latter would be a task requiring only time for its
execution, but requiring more of it than I have now at
my command.

SUMMARY and CONCLUSION.

My desire in these lectures has been to show you, with as little breach of continuity as possible, something of the past growth and present aspect of a department of science, in which have laboured some of the greatest intellects the world has ever seen. My friend Professor Henry, in introducing me at Washington, spoke of me as an apostle; but the only apostolate that I intended to fulfil was to place, in plain words, my subject before you, and to permit its own intrinsic attractions to act upon your minds. I have sought to confer upon each experiment a distinct intellectual value, for experiments ought to be the representatives and expositors of thought—a language addressed to the eye as spoken words are to the ear. In association with its context, nothing is more impressive or instructive than a fit experiment; but, apart from its context, it rather suits the conjuror's purpose of surprise than that purpose of education which ought to be the ruling motive of the scientific man.

And now a brief summary of our work will not be out of place. Our present mastery over the laws and phenomena of light has its origin in the desire of man to *know*. We have seen the ancients busy with this problem, but, like a child who uses his arms aimlessly for want of the necessary muscular exercise, so these early men speculated vaguely and confusedly regarding light, not having as yet the discipline needed to give clearness to their insight, and firmness to their grasp of principles. They assured themselves of the rectilineal propagation of light, and that the angle of incidence was equal to the angle of reflection. For more than a thousand years—I might say, indeed, for more than fifteen hundred years subsequently—the scientific intellect appears as if smitten with paralysis, the fact being that, during this time, the mental force, which might have run in the direction of science, was diverted into other directions.

The course of investigation as regards light was resumed in 1100 by an Arabian philosopher named Alhazan. Then it was taken up in succession by Roger Bacon, Vitellio, and Kepler. These men, though failing to detect the principle which ruled the facts, kept the fire of investigation constantly burning. Then came the fundamental discovery of Snell, that cornerstone of optics, as I have already called it, and immediately afterwards we have the application by Descartes of Snell's discovery to the explanation of the rainbow. Following this we have the overthrow, by Roemer, of the notion of Descartes that light was transmitted instantaneously through space. Then came Newton's crowning experiments on the analysis and synthesis of white light, by which it was proved to be compounded of various kinds of light of different degrees of refrangibility.

Up to his demonstration of the composition of white light, Newton had been everywhere triumphant— triumphant in the heavens, triumphant on the earth, and his subsequent experimental work is for the most part of immortal value. But infallibility is not the gift of man, and, soon after his discovery of the nature of white light, Newton proved himself human. He supposed that refraction and dispersion went hand in hand, and that you could not abolish the one without at the same time abolishing the other. Here Dolland corrected him.

But Newton committed a graver error than this. Science, as I sought to make clear to you in our second lecture, is only in part a thing of the senses. The roots of phenomena are embedded in a region beyond the reach of the senses, and less than the root of the matter will never satisfy the scientific mind. We find, accordingly, in this career of optics the greatest minds constantly yearning to break the bounds of the senses, and to trace phenomena to their subsensible foundations. Thus impelled they entered the region of theory, and here Newton, though drawn from time to time towards the truth, was drawn still more strongly towards the error, and made it his substantial choice. His experiments are imperishable, but his theory has passed away. For a century it stood like a dam across the course of discovery; but, like all barriers that rest upon authority, and not upon truth, the pressure from behind increased, and eventually swept the barrier away. This, as you know, was done mainly through the labours of Thomas Young, and his illustrious French fellow-worker Fresnel.

In 1808 Malus, looking through Iceland spar at the sun reflected from the window of the Luxembourg Palace in Paris, discovered the polarization of light by reflection. In 1811 Arago discovered the splendid chromatic phenomena which we have had illustrated by plates of gypsum in polarized light; he also dis-

covered the rotation of the plane of polarization by
quartz-crystals. In 1813 Seebeck discovered the polarization of light by tourmaline. That same year Brewster discovered those magnificent bands of colour that
surround the axes of biaxal crystals. In 1814 Wollaston discovered the rings of Iceland spar. All these
effects, which without a theoretic clue would leave the
human mind in a jungle of phenomena without harmony or relation, were organically connected by the
theory of undulation.

The theory was applied and verified in all directions, Airy being especially conspicuous for the severity
and conclusiveness of his proofs. The most remarkable verification fell to the lot of the late Sir William
Hamilton, of Dublin, who, taking up the theory
where Fresnel had left it, arrived at the conclusion
that at four special points at the surface of the
ether-wave in double-refracting crystals the ray was
divided not into two parts, but into an infinite
number of parts; forming at these points a continuous conical envelope instead of two images. No
human eye had ever seen this envelope when Sir
William Hamilton inferred its existence. He asked
Dr. Lloyd to test experimentally the truth of his
theoretic conclusion. Lloyd, taking a crystal of arra-
gonite, and following with the most scrupulous exactness the indications of theory, cutting the crystal where
theory said it ought to be cut, observing it where
theory said it ought to be observed, discovered the
luminous envelope which had previously been a mere
idea in the mind of the mathematician.

Nevertheless this great theory of undulation, like
many another truth, which in the long run has proved

a blessing to humanity, had to establish, by hot conflict, its right to existence. Great names were arrayed against it. It had been enunciated by Hooke, it had been applied by Huyghens, it had been defended by Euler. But they made no impression. And, indeed, the theory in their hands was more an analogy than a demonstration. It first took the form of a demonstrated verity in the hands of Thomas Young. He brought the waves of light to bear upon each other, causing them to support each other, and to extinguish each other at will. From their mutual actions he determined their lengths, and applied his determinations in all directions. He showed that the standing difficulty of polarization might be embraced by the theory.

After him came Fresnel, whose transcendent mathematical abilities enabled him to give the theory a generality unattained by Young. He grasped the theory in its entirety; followed the ether into the hearts of crystals of the most complicated structure, and into bodies subjected to strain and pressure. He showed that the facts discovered by Malus, Arago, Brewster, and Biot were so many ganglia, so to speak, of his theoretic organism, deriving from it sustenance and explanation. With a mind too strong for the body with which it was associated, that body became a wreck long before it had become old, and Fresnel died, leaving, however, behind him a name immortal in the annals of science.

One word more I should like to say regarding Fresnel. There are things, ladies and gentlemen, better even than science. Character is higher than Intellect, but it is especially pleasant to those who wish to think well of human nature when high intellect and upright character are combined. They were, I believe, combined in this young Frenchman. In those hot conflicts of the undulatory theory, he stood forth as a man of integrity, claiming no more than his right, and ready to concede their rights to others. He at once recognized and acknowledged the merits of Thomas Young. Indeed, it was he, and his fellow-countryman Arago, who first startled England into the consciousness of the injustice done to Young in the *Edinburgh Review.*

I should like to read you a brief extract from a letter written by Fresnel to Young in 1824, as it throws a pleasant light upon the character of the French philosopher. ' For a long time,' says Fresnel, ' that sensibility, or that vanity, which people call love of glory has been much blunted in me. I labour much less to catch the suffrages of the public than to obtain that inward approval which has always been the sweetest reward of my efforts. Without doubt, in moments of disgust and discouragement, I have often needed the spur of vanity to excite me to pursue my researches. But all the compliments I have received from Arago, De la Place, and Biot never gave me so much pleasure as the discovery of a theoretic truth, or the confirmation of a calculation by experiment.'

This, ladies and gentlemen, is the core of the whole matter as regards science. It must be cultivated for its own sake, for the pure love of truth, rather than for the applause or profit that it brings. And now my occupation in America is well-nigh gone. Still I will bespeak your tolerance for a few concluding remarks in reference to the men who have bequeathed to us the vast body of knowledge of which I have sought to give you some faint idea in these lectures. What was the motive that spurred them on ? What urged them to those battles and those victories over reticent Nature which have become the heritage of the human race ? It is never to be forgotten that not one of those great investigators, from Aristotle down to Stokes and Kirchhoff, had any practical end in view, according to the ordinary definition of the word 'practical.' They did not propose to themselves money as an end, and knowledge as a means of obtaining it. For the most part, they nobly reversed this process, made knowledge their end, and such money as they possessed the means of obtaining it.

We may see to-day the issues of their work in a thousand practical forms, and this may be thought sufficient to justify, if not ennoble their efforts. But they did not work for such issues; their reward was of a totally different kind. In what way different ? We love clothes, we love luxuries, we love fine equipages, we love money, and any man who can point to these as the result of his efforts in life justifies these results before all the world. In America and England more especially he is a 'practical' man. But I would appeal confidently to this assembly whether such things exhaust the demands of human nature ? The very presence here for six inclement nights of this audience, embodying so much of the mental force and refinement of this great city, is an answer to my question. I need not tell such an assembly that there are joys of the intellect as well as joys of the body, or that these pleasures of the spirit constituted the reward of our great investigators. Led on by the whisperings of natural truth, through pain and self-denial, they often pursued their work. With

the ruling passion strong in death, some of them, when
no longer able to hold a pen, dictated to their friends
the results of their labours, and then rested from them
for ever.

Could we have seen these men at work without any
knowledge of the consequences of their work, what
should we have thought of them? To the uninitiated,
in their day, they might often appear as big children
playing with soap-bubbles and other trifles. It is
so to this hour. Could you watch the true investigator—your Henry or your Draper, for example—in
his laboratory, unless animated by his spirit, you could
hardly understand what keeps him there. Many of
the objects which rivet his attention might appear to
you utterly trivial; and, if you were to ask him
what is the *use* of his work, the chances are that
you would confound him. He might not be able
to express the use of it in intelligible terms. He
might not be able to assure you that it will put a
dollar into the pocket of any human being living or to
come. That scientific discovery *may* put not only
dollars into the pockets of individuals, but millions
into the exchequers of nations, the history of science
amply proves; but the hope of its doing so never was
and never can be the motive power of the investigator.

I know that I run some risk in speaking thus before
practical men. I know what De Tocqueville says of
you. 'The man of the North,' he says, 'has not only
experience, but knowledge. He, however, does not
care for science as a pleasure, and only embraces it
with avidity when it leads to useful applications.' But
what, I would ask, are the hopes of useful applications
which have caused you so many times to fill this place

in spite of snow-drifts and biting cold ?   What, I may ask, is the origin of that kindness which drew me from my work in London to address you here, and which, if I permitted it, would send me home a millionaire ? Not because I had taught. you to make a single cent by science am. I here to-night, but because I tried to the best of my ability to present science to the world as an intellectual good.   Surely no two terms were ever so distorted and misapplied with reference to man in his higher relations as these terms useful and practical. Let us expand their definitions until they embrace all the needs of man, his highest intellectual needs inclu- sive.   It is specially on this ground of its adminis- tering to the higher needs of the intellect ; it is mainly because I believe it to be wholesome, both as a source of knowledge and as a means of discipline, that I urge the claims of science upon your attention.

But with reference to material needs and joys, surely pure science has also a word to say.   People sometimes speak as if steam had not been studied before James Watt, or electricity before Wheatstone and Morse ; whereas, in point of fact, Watt and Wheatstone and Morse, with all their practicality, were the mere out- come of antecedent forces, which acted without refer- ence to practical ends.   This also, I think, merits a moment's attention.   You are delighted, and with good reason, with your electric telegraphs, proud of your steam-engines and your factories, and charmed with the productions of photography.   You see daily, with just elation, the creation of new forms of industry— new powers of adding to the wealth and comfort of society.   Industrial England is heaving with forces tending to this end, and the pulse of industry beats

still stronger in the United States. And yet, when
analyzed, what are industrial America and industrial
England ?

If you can tolerate freedom of speech on my part,
I will answer this question by an illustration. Strip
a strong arm, and regard the knotted muscles when
the hand is clenched and the arm bent. Is this
exhibition of energy the work of the muscle alone?
By no means. The muscle is the channel of an influence, without which it would be as powerless as a lump
of plastic dough. It is the delicate unseen nerve that
unlocks the power of the muscle. And without those
filaments of genius which have been shot like nerves
through the body of society by the original discoverer,
industrial America and industrial England would be
very much in the condition of that plastic dough.

At the present time there is a cry in England for
technical education, and it is a cry in which the most
commonplace intellect can join, its necessity is so
obvious. But there is no cry for original investigation. Still without this, as surely as the stream
dwindles when the spring dies, so surely will ' technical
education ' lose all force of growth, all power of reproduction. Our great investigators have given us
sufficient work for a time ; but if their spirit die out,
we shall find ourselves eventually in the condition of
those Chinese mentioned by De Tocqueville, who,
having forgotten the scientific origin of what they
did, were at length compelled to copy without variation the inventions of an ancestry who, wiser than
themselves, had drawn their inspiration direct from
Nature.

Both England and America have reason to bear those

things in mind, for the largeness and nearness of material results are only too likely to cause both countries to forget the small spiritual beginnings of such results in the mind of the scientific discoverer. You multiply, but he creates. And if you starve him, or otherwise kill him—nay, if you fail to secure for him free scope and encouragement—you not only lose the motive power of intellectual progress, but infallibly sever yourselves from the springs of industrial life.

What has been said of technical operations holds equally good for education, for here also the original investigator constitutes the fountain-head of knowledge. It belongs to the teacher to give this knowledge the requisite form ; an honourable and often a difficult task. But it is a task which receives its final sanctification when the teacher himself honestly tries to add a rill to the great stream of scientific discovery. Indeed, it may be doubted whether the real life of science can be fully felt and communicated by the man who has not himself been taught by direct communion with Nature. We may, it is true, have good and instructive lectures from men of ability, the whole of whose knowledge is second-hand, just as we may have good and instructive sermons from intellectually able and unregenerate men. But for that power of science which corresponds to what the Puritan fathers would call experimental religion in the heart, you must ascend to the original investigator.

To keep society as regards science in healthy play, three classes of workers are necessary : Firstly, the investigator of natural truth, whose vocation it is to pursue that truth, and extend the field of discovery for the truth's own sake, and without reference to practical ends. Secondly, the teacher of natural truth, whose vocation it is to give public diffusion to the knowledge already won by the discoverer. Thirdly, the applier of natural truth, whose vocation it is to make scientific knowledge available for the needs, comforts, and luxuries of life. These three classes ought to co-exist and interact. Now, the popular notion of science, both in this country and in England, often relates not to science strictly so called, but to the applications of science. Such applications, especially on this continent, are so astounding—they spread themselves so largely and umbrageously before the public eye—as to shut out from view those workers who are engaged in the quieter and profounder business of original investigation.

Take the electric telegraph as an example, which has been repeatedly forced upon my attention of late. I am not here to attenuate in the slightest degree the services of those who, in England and America, have given the telegraph a form so wonderfully fitted for public use. They earned a great reward, and assuredly they have received it. But I should be untrue to you and to myself if I failed to tell you that, however high in particular respects their claims and qualities may be, your practical men did not discover the electric telegraph. The discovery of the electric telegraph implies the discovery of electricity itself, and the development of its laws and phenomena. Such discoveries are not made by practical men, and they never will be made by them, because their minds are beset by ideas which, though of the highest value from one point of view, are not those which stimulate the original discoverer.

The ancients discovered the electricity of amber ; and Gilbert, in the year 1600, extended the discovery to other bodies. Then followed other inquirers, your own Franklin among the number. But this form of electricity, though tried, did not come into use for telegraphic purposes. Then appeared the great Italian Volta, who discovered the source of electricity which bears his name, and applied the most profound insight and the most delicate experimental skill to its development. Then arose the man who added to the powers of his intellect all the graces of the human heart, Michael Faraday, the discoverer of the great domain of magneto-electricity. Œrsted discovered the deflection of the magnetic needle, and Arago and Sturgeon the magnetization of iron by the electric current. The voltaic circuit finally found its theoretic Newton in Ohm, while Henry, of Princeton, who had the sagacity to recognize the merits of Ohm while they were still decried in his own country, was at this time in the van of experimental inquiry.

In the works of these men you have all the materials employed at this hour in all the forms of the electric telegraph. Nay, more ; Gauss, the celebrated astronomer, and Weber, the celebrated natural philosopher, both professors in the University of Göttingen, wishing to establish a rapid mode of communication between the observatory and the physical cabinet of the university, did this by means of an electric telegraph. Thus, before your practical men appeared upon the scene, the force had been discovered, its laws investigated and made sure, the most complete mastery of its phenomena had been attained—nay, its applicability to telegraphic purposes demonstrated—by men whose sole

reward for their labours was the noble excitement of
research and the joy attendant on the discovery of
natural truth.

Are we to ignore all this ? We do so at our peril.
For I say again that, behind all our practical applications, there is a region of intellectual action to which
practical men have rarely contributed, but from which
they draw all their supplies. Cut them off from this
region, and they become eventually helpless.   In no case
is the adage truer, ' Other men laboured, but ye are
entered into their labours,' than in the case of the discoverer and applier of natural truth. But now a word
on the other side. While I say that practical men are
not the men to make the necessary antecedent discoveries, the cases are rare in which the discoverer
knows how to turn his labours to practical account.
Different qualities of mind and different habits of
thought are needed in the two cases ; and while I wish
to give emphatic utterance to the claims of those whose
position, owing to the simple fact of their intellectual
elevation, is often misunderstood, I am not here to
exalt the one class of workers at the expense of the
other.   They are the necessary complements of each
other. But remember that one class is sure to be taken
care of. All the material rewards of society are already
within their reach, while that same society habitually
ascribes to them intellectual achievements which were
never theirs. This cannot but act to the detriment of
those profounder studies out of which, not only our
knowledge of nature, but our present industrial arts
themselves have sprung, and from which the rising
genius of the country is incessantly tempted away.

Pasteur, one of the most eminent members of the

Institute of France, in accounting for the disastrous overthrow of his country and the predominance of Germany in the late war, expresses himself thus : 'Few persons comprehend the real origin of the marvels of industry and the wealth of nations. I need no further proof of this than the employment more and more frequent in official language, and in writing of all sorts, of the erroneous expression *applied science.* The abandonment of scientific careers by men capable of pursuing them with distinction was recently deplored in the presence of a minister of the greatest talent. This statesman endeavoured to show that we ought not to be surprised at this result, because *in our day the reign of theoretic science yielded place to that of applied science.* Nothing could be more erroneous than this opinion, nothing, I venture to say, more dangerous, even to practical life, than the consequences which might flow from these words. They have rested on my mind as a proof of the imperious necessity of reform in our superior education. There exists no category of the sciences to which the name of applied science could rightly be given. *We have science, and the applications of science,* which are united together as the tree and its fruit.'

And Cuvier, the great comparative anatomist, writes thus upon the same theme : 'These grand practical innovations are the mere applications of truths of a higher order, not sought with a practical intent, but which were pursued for their own sake, and solely through an ardour for knowledge. Those who applied them could not have discovered them ; those who discovered them had no inclination to pursue them to a practical end. Engaged in the high regions whither

their thoughts had carried them, they hardly perceived
these practical issues, though born of their own deeds.
These rising workshops, these peopled colonies, those
ships which furrow the seas—this abundance, this
luxury, this tumult—all this comes from discoverers in
science, and it all remains strange to them. At the
point where science merges into practice they abandon
it; it concerns them no more.'

When the Pilgrim Fathers landed at Plymouth
Rock, and when Penn made his treaty with the Indians,
the new-comers had to build their houses, to chasten
the earth into cultivation, and to take care of their
souls. In such a community science, in its more abstract forms, was not to be thought of. And at the
present hour, when your hardy Western pioneers stand
face to face with stubborn Nature, piercing the mountains and subduing the forest and the prairie, the pursuit of science for its own sake is not to be expected.
The first need of man is food and shelter; but a vast,
portion of this continent is already raised far beyond
this need. The gentlemen of New York, Brooklyn,
Boston, Philadelphia, Baltimore, and Washington have
already built their houses, and very beautiful they are;
they have also secured their dinners, to the excellence
of which I can also bear testimony. They have, in
fact, reached that precise condition of well-being and
independence when a culture, as high as humanity has
yet reached, may be justly demanded at their hands.
They have reached that maturity, as possessors of
wealth and leisure, when the investigator of natural
truth, for the truth's own sake, ought to find among
them promoters and protectors.

Among the many problems before them they have

this to solve, whether a republic is able to foster the highest forms of genius. You are familiar with the writings of De Tocqueville, and must be aware of the intense sympathy which he felt for your institutions; and this sympathy is all the more valuable from the philosophic candour with which he points out not only your merits, but your defects and dangers. Now if I come here to speak of science in America in a critical and captious spirit, an invisible radiation from my words and manner will enable you to find me out, and will guide your treatment of me to-night. But if I in no unfriendly spirit—in a spirit, indeed, the reverse of unfriendly—venture to repeat before you what this great historian and analyst of democratic institutions said of America, I am persuaded that you will hear me out. He wrote some three-and-twenty years ago, and perhaps would not write the same to-day; but it will do nobody any harm to have his words repeated, and, if necessary, laid to heart.

In a work published in 1850, De Tocqueville says: 'It must be confessed that, among the civilized peoples of our age, there are few in which the highest sciences have made so little progress as in the United States.'[1] He declares his conviction that, had you been alone in the universe, you would speedily have discovered that you cannot long make progress in practical science without cultivating theoretic science at the same time. But, according to De Tocqueville, you are not thus alone. He refuses to separate America from its ancestral home; and it is

[1] 'Il faut reconnaître que parmi les peuples civilisés de nos jours il en est peu chez qui les hautes sciences aient fait moins de progrès qu'aux États-Unis, ou qui aient fourni moins de grands artistes, de poètes illustres et de célèbres écrivains.' (De la Démocratie en Amérique, etc., tome ii. p. 36.)

here, he contends, that you collect the treasures of the intellect, without taking the trouble to create them.

De Tocqueville evidently doubts the capacity of a democracy to foster genius as it was fostered in the ancient aristocracies. 'The future,' he says, 'will prove whether the passion for profound knowledge, so rare and so fruitful, can be born and developed so readily in democratic societies as in aristocracies. As for me,' he continues, 'I can hardly believe it.' He speaks of the unquiet feverishness of democratic communities, not in times of great excitement, for such times may give an extraordinary impetus to ideas, but in times of peace. There is then, he says, 'a small and uncomfortable agitation, a sort of incessant attrition of man against man, which troubles and distracts the mind without imparting to it either loftiness or animation.' It rests with you to prove whether these things are necessarily so—whether the highest scientific genius cannot find in the midst of you a tranquil home.

I should be loth to gainsay so keen an observer and so profound a political writer, but, since my arrival in this country, I have been unable to see anything in the constitution of society to prevent a student with the root of the matter in him from bestowing the most steadfast devotion on pure science. If great scientific results are not achieved in America, it is not to the small agitations of society that I should be disposed to ascribe the defect, but to the fact that the men among you who possess the endowments necessary for profound scientific inquiry are laden with duties of administration or tuition so heavy as to be utterly incompatible with the continuous and tranquil meditation which original investigation demands. It may well be asked whether Henry would have been transformed into an administrator, or whether Draper would have forsaken science to write history, if the original investigator had been honoured as he ought to be in this land. I hardly think they would. Still I do not imagine this state of things likely to last. In America there is a willingness on the part of individuals to devote their fortunes in the matter of education to the service of the commonwealth, which is probably without a parallel elsewhere; and this willingness requires but wise direction to enable you effectually to wipe away the reproach of De Tocqueville.

Your most difficult problem will be not to build institutions, but to discover men. You may erect laboratories and endow them; you may furnish them with all the appliances needed for enquiry; in so doing you are but creating opportunity for the exercise of powers which come from sources entirely beyond your reach. You cannot create genius by bidding for it. In biblical language, it is the gift of God; and the most you could do, were your wealth, and your willingness to apply it, a million-fold what they are, would be to make sure that this glorious plant shall have the freedom, light, and warmth, necessary for its development. We see from time to time a noble tree dragged down by parasitic runners. These the gardener can remove, though the vital force of the tree itself may lie beyond him; and so, in many a case, you men of wealth can liberate genius from the hampering toils which the struggle for existence often casts around it.

Drawn by your kindness, I have come here to give these lectures, and now that my visit to America has become almost a thing of the past, I look back upon it as a memory without a single stain. No lecturer was

Q

ever rewarded as I have been. From this vantage-
ground, however, let me remind you that the work of
the lecturer is not the highest work; that in science
the lecturer is usually the distributor of intellectual
wealth amassed by better men. And though lecturing
and teaching, in moderation, will in general promote
their moral health, it is not solely, or even chiefly, as
lecturers, but as investigators, that your highest men
ought to be employed. You have scientific genius
amongst you—not sown broadcast, believe me, it is sown
thus nowhere—but still scattered here and there.
Take all unnecessary impediments out of its way.
Keep your sympathetic eye upon the originator of
knowledge. Give him the freedom necessary for his
researches, not overloading him either with the duties
of tuition or of administration, not demanding from
him so-called practical results—above all things,
avoiding that question which ignorance so often ad-
dresses to genius, 'What is the use of your work?'
Let him make truth his object, however unpractical
for the time being that truth may appear. If you
cast your bread thus upon the waters, then be assured
it will return to you, though it may be after many days.

APPENDIX.

———◇———

LORD BROUGHAM'S ARTICLES ON DR. THOMAS YOUNG IN THE 'EDINBURGH REVIEW.'

IN Lecture II. of the foregoing series, the attacks of the Edinburgh Reviewers on the scientific labours of Dr. Young are briefly referred to. The spirit of these attacks will be understood from the extracts given below. They had, it is to be feared, a very damaging effect, both upon Young's reputation, and upon his scientific activity. The first of them, published in No. II. of the Review, was levelled at Young's memoir on the Theory of Light and Colours, which had been chosen by the Royal Society as the Bakerian Lecture for 1801.

'As this paper,' says the Reviewer, 'contains nothing which deserves the name, either of experiment or discovery, and as it is in fact destitute of every species of merit, we should have allowed it to pass among the multitude of those articles which must always find admittance into the collections of a Society which is pledged to publish two or three volumes every year. The dignities of the author, and the title of Bakerian Lecture, which is prefixed to these lucubrations, should not have saved them from a place in the ignoble crowd. But we have of late observed in the physical world a most unaccountable predilection for vague hypothesis daily gaining ground; and we are mortified to see that the Royal Society, forgetful of those improvements in science to which it owes its origin, and neglecting the precepts of its most illustrious members, is now, by the publication of such papers, giving the countenance of its highest authority to dangerous relaxations in the principles of physical logic. We wish to raise our feeble voice against innovations that can have no other effect than to check the progress of Science, and renew all those wild phantoms of the imagination which Bacon and Newton put to flight from her temple. We wish to recall philosophers to the strict and severe methods of investigation pointed out by the transcendent talents of those illustrious men, and consecrated by their astonishing success; and, for this purpose, we take the first opportunity that has been presented to us of calling our readers' attention to this mode of philosophising, which seems, by the title of the paper now before us, to have been honoured with more than the ordinary approbation of the Council. . . . .

'It is difficult to deal with an author whose mind is filled with a medium of so fickle and vibratory a nature. Were we to take the trouble of refuting him, he might tell us, "*My opinion is changed*, and *I have abandoned that hypothesis, but here is another for you.*" We demand if the world of science which Newton once illuminated is to be as changeable in its modes as the world of fashion, which is directed by the nod of a silly woman or a pampered fop? Has the Royal Society degraded its publications into bulletins of new and fashionable theories for the ladies of the Royal Institution? *Proh pudor!* Let the Professor continue to amuse his audience with an endless variety of such harmless trifles, but, in the name of science, let them not find admittance into that venerable repository which contains the works of Newton, and Boyle, and Cavendish, and Maskelyne, and Herschel.

'These remarks lead us to observe, that perpetual fluctuation and change of ground is the common lot of theorists. An hypothesis which is assumed from a fanciful analogy or adopted from its apparent capacity of explaining certain appearances, must always be varied as new facts occur, and must be kept alive by a repetition of the same process of touching and retouching, of successive accommodation and adaptation, to which it originally owed its puny and contemptible existence. But the making of an hypothesis is not the discovery of a truth. It is a mere sporting with the subject; it is a sham fight which may amuse in the moment of idleness and relaxation, but will neither gain victories over prejudice and error, nor extend the empire of science. A mere theory is in truth destitute of merit of every kind, except that of a warm and misguided imagination. It demonstrates neither patience of investigation, nor rich resources of skill, nor vigorous habits of attention, nor powers of abstracting and comparing, nor extensive acquaintance with nature. It is the unmanly and unfruitful pleasure of a boyish prurient imagination, or the gratification of a corrupted and depraved appetite.

'If, however, we condescend to amuse ourselves in this manner we have a right to demand that the entertainment shall be of the right sort, and that the hypothesis shall be so consistent with it-self, and so applicable to the facts, as not to require perpetual mending and patching; that the child that we stoop to play with shall be tolerably healthy, and not of the puny, sickly nature of Dr. Young's productions, which have scarcely *stamina* to subsist until the fruitful parent has furnished us with a new litter, to make way for which he knocks on the head or more barbarously exposes the first. . . . .

' A little further acquaintance, however, with the Doctor's paper has convinced us that he is as little scrupulous in his quotations as in his theories; that he delights as much to twist an authority as to torture a fact; and according to his usual vibratory method, after a second examination of the Newtonian writings, has changed the opinion which his first perusal gave him of their significa-tion. . . . .

' After all, it may be said Newton amused himself with hypo-theses, and so may Dr. Young. Admitting that the Doctor's relaxations were the same with his predecessor's, it must be remembered that the queries of Newton were given to the world at the close of the most brilliant career of solid discovery that any mortal was ever permitted to run. The sports in which such a veteran might well be allowed to relax his mind, are mere idleness in the raw soldier who has never fleshed his sword; and though the world would gaze with interest upon every such occupation of the former, they would turn with disgust from the forward and idle attempts of the latter to obtrude upon them his awkward gambols. . . . .

' From such a dull invention (the Ether) nothing can be expected. . . . It teaches no truth, reconciles no contradictions, arranges no anomalous facts, suggests no new experiments, and leads to [no] new inquiries. It has not even the pitiful merit of affording an agreeable play to the fancy. It is infinitely more useless, and less ingenious, than the Indian theory of the Elephant and Tortoise. It may be ranked in the same class with that stupid invention of metaphysical theology, &c.'

The first volume of the Review contains a second article by the same hand, attacking Young's paper en-titled ' An Account of some Cases of the Production of Colours not hitherto described,' published in the ' Philosophical Transactions ' for 1802. Here is a sample of the style in which the Reviewer handles this paper :—

' We cannot conclude our review of these articles without en-treating, for a moment, the attention of that illustrious body which has admitted of late years so many paltry and unsubstantial papers into its 'Transactions.' Great as the services are which the Royal Society has rendered to the world, and valuable as the papers have been in every volume (not less valuable, surely, since the accession of the present excellent President), we think on the benefits which it has conferred with feelings of the warmest gratitude. We only wish that those feelings should be unmingled by any ideas of regret, from the want of selection to which we are adverting; and that it should cease to give its countenance to such vain theories as those which we find mingled, in this volume, with a vast body of important information. The Society has, indeed, been in the habit of stating that the truth and other merits of the speculations which it publishes must rest with their respective authors; but we are afraid this is not sufficient. The Society publishes these papers—meets for the purpose of reading them—calls them its 'Transactions;' and, in fact, exercises, in many cases, the power of rejecting the papers which are offered. It is in vain, therefore, to disavow a responsibility which so many circumstances concur in fixing. The public will always look upon the Society as immediately responsible for the papers which compose its 'Transactions,' unless, indeed, it shall wish to be degraded into the rank of a mere mechanical contrivance for the printing of miscellanies. We implore the Council, therefore, if they will deign to cast their eyes upon our humble page, to prevent a degradation of the Institution which has so long held the first rank among scientific bodies. Let them reflect on the mighty name which has been transmitted to them—

———— 'Clarum et venerabile nomen
Gentibus, et multum nostræ quod proderat urbi.'

Such a name may indeed shelter them in their weakness, and make us venerate, even in the frailty of old age, an institution illustrious for its ancient virtue. But is it impossible to ward off the encroachments of time, and to renovate, in new achievements, the vigour of former years? It is more honourable to support an illustrious character, than to appeal to it for shelter and protection.'

In Vol. V. of the Review we have a criticism of Young's paper entitled 'Experiments and Calculations relative to Physical Optics,' which was chosen by the Royal Society as the Bakerian Lecture for 1804.

'On a former occasion we addressed some remarks to the author of this paper, and took the liberty also of offering a few humble suggestions to the illustrious Body in whose memoirs it is published. The long silence which he has since preserved on philosophical matters, at least through this channel of communication with the scientific world, led us to flatter ourselves either that he had discontinued his fruitless chase after hypotheses, or that the Society had remitted his effusions to the more appropriate audience of both sexes which throngs round the chairs of the Royal Institution. The volume now before us, however, at once destroys such expectations. The paper which stands first is another Bakerian Lecture, containing more fancies, more blunders, more unfounded hypotheses, more gratuitous fictions, all upon the same field on which Newton trode, and all from the fertile, yet fruitless, brain of the same eternal Dr. Young.'

The Reviewer thus winds up the controversy :—

'We now dismiss, for the present, the feeble lucubrations of this author, in which we have searched without success for some traces of learning, acuteness, and ingenuity, that might compensate his evident deficiency in the powers of solid thinking, calm and patient investigation, and successful development of the laws of Nature, by steady and modest observation of her operations. We came to the examination with no other prejudice than the very allowable prepossession against vague hypothesis, by which all true lovers of science have for above a century and a half been swayed. We pursued it, both on the present and on a former occasion, without any feelings except those of regret at the abuse of that time and opportunity which no greater share of talents than Dr. Young's are sufficient to render fruitful by mere diligence and moderation. From us, however, he cannot claim any portion of respect until he shall alter his mode of proceeding, or change the subject of his lucubrations; and we feel ourselves more particularly called upon to express our disapprobation because, as distinction has been unwarily bestowed on his labours by the most illustrious of scientific bodies, it is the more necessary that a free protest should be recorded before the more humble tribunals of literature.'

In Lecture II. the possible effect of these attacks upon Young's productiveness as an investigator is referred to. The Reviewer here glances at the silence which he regarded as the result of his invective, with evident satisfaction. It is now time to show how Young met these assaults. Here is his reply to the Edinburgh Reviewers :—

DR. YOUNG'S REPLY TO THE ANIMADVER-
SIONS OF THE EDINBURGH REVIEWERS.

A man who has a proper regard for the dignity of his own
character, although his sensibility may sometimes be awakened
by the unjust attacks of interested malevolence, will esteem it
in general more advisable to bear, in silence, the temporary
effects of a short-lived injury, than to suffer his own pursuits to
be interrupted, in making an effort to repel the invective, and
to punish the aggressor. But it is possible that art and malice
may be so insidiously combined, as to give to the grossest misrepresentations the semblance of justice and candour; and,
especially where the subject of the discussion is of a nature
little adapted to the comprehension of the generality of readers,
even a man's friends may be so far misled by a garbled extract
from his own works, and by the specious mixture of partial
truth with essential falsehood, that they may not only be unable
to defend him from the unfavourable opinion of others, but may
themselves be disposed to suspect, in spite of their partiality,
that he has been hasty and inconsiderate at least, if not
radically weak and mistaken. In such a case, he owes to his
friends such explanations as will enable them to see clearly the
injustice of the accusation, and the iniquity of its author: and,
if he is in a situation which requires that he should in a certain
degree possess the public confidence, he owes to himself and to
the public to prove that the charges of imbecility of mind and
perversity of disposition are not more founded with regard to
him, than with regard to all who are partakers with him in the
unavoidable imperfections of human nature.

Precisely such is my situation. I have at various times
communicated to the Royal Society, in a very abridged form,
the results of my experiments and investigations relating to
different branches of natural philosophy: and the Council of the
Society, with a view perhaps of encouraging patient diligence,
has honoured my essays with a place in their 'Transactions.'
Several of these essays have been singled out, in an unprece-

dented manner, from the volumes in which they were printed, and have been made the subjects, in the second and ninth numbers of the 'Edinburgh Review,' not of criticism, but of ridicule and invective; of an attack not only upon my writings and my literary pursuits, but almost on my moral character. The peculiarity of the style and tendency of this attack led me at once to suspect that it must have been suggested by some other motive than the love of truth; and I have both internal and external evidence for believing that the articles in question are, either wholly or in great measure, the productions of an individual upon whose mathematical works I had formerly thought it necessary to make some remarks, which, though not favourable, were far from being severe;[1] and whose optical speculations, partly confuted before, and already forgotten, appeared, to their fond parent, to be in danger of a still more complete rejection from the establishment of my opinions. As far as my reputation in natural philosophy is concerned, I should consider a libel of this kind as neither requiring nor deserving an answer; but I cannot help feeling the propriety of endeavouring to defend myself from the more pernicious influence of those imputations, which might tend to lessen the confidence of the public in the professional qualifications of a man whose abilities have been thus contemptuously and repeatedly depreciated. The practice of physic has always been, either immediately or remotely, the object of my pursuits, and I can affirm, without fear of contradiction, that I have never neglected any opportunity either of improving myself in its study, or of being useful to the humblest of those who have committed themselves to my care in its application. But I have no right to expect that any degree of industry that I may have employed, should encourage a man to entrust me with the management of that which so nearly concerns his happiness and prosperity, if he has reason to think me rash, and vain, and wavering in my opinions, and that even upon subjects which are

---

[1] Young's *Miscellaneous Works*, vol. i. p. 101; see also note at the foot of p. 99.—*Note by Dean Peacock.*

generally supposed to admit of proofs perfectly decisive and satisfactory.

My Bakerian lecture on the theory of light and colours, and another paper published in the same volume of the 'Philosophical Transactions,' are the subjects of two of the most scurrilous articles in the second number of the 'Edinburgh Review.' The writer of these articles has, as a prelude to his imputation of a 'vibratory and undulatory mode of reasoning,' very unnecessarily recurred to the first essay that I presented to the Royal Society, as long ago as the year 1793; I am therefore obliged to explain the circumstances which led me to the subject of that essay, and to relate the history of my opinions concerning it: and as he has thought proper to insinuate, in the form of insolent admonition, that I have never studied even 'the plainer parts' of the works of Newton, I must state when and why I actually read those admirable productions; and I shall think it right to account, at the same time, for the manner in which, as a medical man, I have been led, for a time, into the extensive regions of natural philosophy.

It is now more than fourteen years since I first resolved to devote my life to the profession of physic. I continued for two years the pursuit of those attainments, in mathematics and in general literature, which had before constituted my sole occupation, and which, by the express sentiment of the father of the medical sciences, and by the universal suffrages of the more liberal part of mankind, have been allowed to be the surest and best foundations for the superstructure of the requisite qualifications of a physician. The causes of disease, obscure in their nature, and hidden in their operation, elude but too frequently the most diligent researches of the strongest and most experienced minds: they afford ample scope to the most minute investigation, and the most sagacious discernment; but they require that the faculties of the observer should have been sufficiently prepared by being employed on subjects of a nature more certainly definable, and more perfectly intelligible. Classical literature, mathematical philosophy, chemistry and natural history, a knowledge of different countries, and an acquaintance with different languages, are as necessary to the melioration of those powers of reasoning which are to be called into activity in the pursuit of a profession, as they are essential to the perfection of the character of a general scholar and an accomplished man. This must be my excuse for having devoted a considerable portion of my attention to the study of the classics, on my success in which the Edinburgh Reviewers have, with an insulting affectation of candour, thought fit, on another occasion, to compliment me. I pursued the study of mathematics and natural philosophy as far only as I esteemed them subservient to other objects: not that I preferred philology to science, but because I thought myself obliged to sacrifice both to physic. After having rendered myself familiar with many other mathematical works, I read, in the autumn of 1790, both the 'Principia' of Newton and his 'Optics.' I read not the 'plaincst parts of the "Principia"' only, but the whole; and all that the illustrious author meant to be understood by a reader, I understood and admired: where he purposely omitted a demonstration, I did not at that time attempt to investigate it. That I was then satisfied with some few parts which I do not now think unexceptionable, might easily have happened, even if I had felt less reverence than I have uniformly done for the character of the unrivalled author. The 'Optics' too I read with attention and delight, yet by no means with the same satisfaction that I had derived from the perusal of the 'Principia.'

My attention to optical subjects was not revived till the year 1793, when, in the course of my anatomical studies, the theory of vision was necessarily to be reconsidered. I saw, what I then thought none had seen before, that the crystalline lens was of a fibrous structure; and I could find no other satisfactory mode of explaining the phenomena of vision than by attributing to it muscular powers. On this subject I presented a short paper to the Royal Society,[1] to which, from the circumstance of the late Mr. Hunter's reclamation of the discovery as his own, a greater degree of novelty was imputed than it perhaps

[1] Young's *Miscellaneous Works*, vol. i. No. I. p. 1.

deserved. Mr. Home too attributed to Mr. Hunter the merit of a discovery 'not small nor unimportant,' that of an animal in which the fibrous structure of the lens was easily traced. I had however found no difficulty in observing, in the eye of a quadruped, the arrangement which had been the basis of my speculations.

It was in the course of the winter which I spent in pursuing my medical studies at Edinburgh, that I first read Mr. Home's account of his experiments on vision.[1] This investigation convinced Mr. Home that Mr. Hunter, whose sentiments he had before adopted, was mistaken in his opinion; and when I had afterwards seen at Göttingen Dr. Olbers' elegant dissertation on the same subject, I found it impossible to resist, without making further experiments of my own, the appearance of evidence which was brought against my favourite opinion. I had not then learned of the Edinburgh Reviewers how much easier it was to deny the accuracy of the experiments of my adversaries than to oppose them by arguments, or to allow due weight to their apparent consequences; and I thought it more honourable to acknowledge my conviction of their importance, than to persist either in error or in silence. I judged, with respect to the matter of fact, perhaps erroneously, but with regard to all the evidence that was then in existence, I judged as every unprejudiced mind must have been inclined to do. It was only in the year 1800 that I was induced to resume the investigation, in consequence of reading, in the medical essays of a society in Edinburgh, Dr. Porterfield's valuable paper ' On the Internal Changes of the Eye.' I improved on his ideas of the construction of an optometer, and I obtained, by numerous and diversified experiments, such accumulated evidence of the truth of my original opinion, that I was obliged to submit to the unexpected necessity of recurring to it once more. Those who have read my paper, not as a modern reviewer reads, but with patience and attention, will not, I imagine, think that any apology is required for this second change of sentiments. I

[1] The Croonian lecture on Muscular Motion. *Phil. Trans.* for 1794, vol. lxxxiv. p. 1.

cannot, however, refuse myself the pleasure of inserting here a passage from a letter which I have lately received from Dr. Olbers, the discoverer of the planet Pallas, the same whose dissertation on vision I have often quoted with applause. 'You may easily suppose,' says Dr. Olbers, 'that your celebrated essay "On the Mechanism of the Eye,"[1] must have interested me very particularly. I saw indeed that it completely refuted my own theory respecting the changes of the eye; but *my object is to discover truth, and not to support my opinion.*' With such a man as Dr. Olbers, my reviewer would say again, as he has said of me, it would be 'difficult to argue : were we to take the trouble of refuting him, he might tell us, *My opinion is changed.*'

I have now, I trust, vindicated myself from the charge of any unwarrantable inconstancy in the changes which my opinions on the subject of vision have undergone. I shall next enter into a similar explanation of my motives for applying myself to the study of the phenomena of sound and light, and of the progress of my ideas respecting their nature. When I took a degree in physic at Göttingen, it was necessary, besides publishing a medical dissertation, to deliver a lecture upon some subject connected with medical studies: and I chose for this the formation of the human voice. A few pages, containing a table of articulate sounds, were printed at the end of my dissertation ' On the Preservative Powers of the Animal Economy; my uncle, Dr. Brocklesby, at the instance of the late most respectable Dr. Heberden, repeatedly urged me to give some further explanation of the subject to which these characters related. When I began the outline of an essay on the human voice, I found myself at a loss for a perfect conception of what sound was, and during the three years that I passed at Emmanuel College, Cambridge, I collected all the information reating to it that I could procure from books, and I made a variety of original experiments on sounds of all kinds, and on the motions of fluids in general. In the course of these inquiries

---

[1] Young's *Miscellaneous Works*, vol. i. No. II.

I learned, to my surprise, how much further our neighbours on
the Continent were advanced in the investigation of the motions of
sounding bodies and of elastic fluids than any of our own countrymen ; and in making some experiments on the production of
sounds, I was so forcibly impressed with the resemblance of the
phenomena that I saw to those of the colours of thin plates,
with which I was already acquainted, that I began to suspect
the existence of a closer analogy between them than I could
before have easily believed. On further reflection and examination my opinion was confirmed, and as I thought I could
place the question in a clearer light than that in which it had
generally been viewed, I was induced to insert my observations
in a paper, which I presented soon after to the Royal Society,
under the name of ' Outlines of Experiments and Inquiries
respecting Sound and Light.'[1]   A determination to confine my
studies as much as possible to physic was my motive for laying
them before the Society in a state of confessed imperfection. I
am not disposed to overrate their value ; the compliment which
was paid to them by an experienced veteran in philosophy, who
wrote the best articles of the ' Encyclopædia Britannica,' is fully
as much as I can flatter myself that they deserve.[2] The motions
of a stream of air, rendered visible by means of smoke, the
diversified rotations of musical chords, the influence of the mode
of agitation on the natural harmonics of strings, the phenomena
of beats, and of grave harmonics, were examined in a manner
which tended to place in a new point of view a subject certainly curious, and not wholly unimportant.

The opinion respecting light, which I first suggested in this
paper as the most probable, was neither the same with Euler's
nor, as the reviewer falsely asserts, in any degree borrowed
from him. It was precisely the theory of Hooke and of Huygens,
with the adoption of some suggestions made by Newton himself
as not in themselves improbable. The only objection which
Newton makes to the hypothesis thus modified, is this : light
could not be propagated, solely by the undulations of a fluid,

---

[1] Young's *Miscellaneous Works*, No. III. p. 64.   [2] Ibid. p. 134.

without spreading almost equally in all directions; and for this assertion he thinks that there is both experiment and demonstration. His arguments from experiment appear to me to have been sufficiently obviated by what Lambert has advanced in the 'Memoirs of Berlin,' and by Professor Robison's remarks on echoes in the 'Encyclopædia,' as well as by many observations which I have myself made, at different times, on the waves of water. The demonstration is attempted in the 'Principia:' to me it appears to be defective; if I am not allowed to be a competent judge, I can quote others, whose authority will not be denied. Euler has been called by some an indifferent philosopher, but he must at least be allowed to have been perfectly capable of judging of mathematical evidence: he had certainly read the 'Principia,' and he utterly denied the conclusiveness of the argument. D'Alembert was a mathematician of acknowledged eminence, and Lalande's approbation of his sentiments must give them additional weight: both these mathematicians assert, as it appears from Lalande's edition of Montucla, that the arguments are so balanced in favour of the different systems of light, that the safest way is to confess 'our utter ignorance of the manner of its propagation.' The celebrated Laplace, in comparing the opinions respecting light, is contented to call the Newtonian doctrine a hypothesis, which, on account of the facility of its application to the phenomena, is extremely probable. If he had considered the undulatory system as demonstrably absurd, he certainly would not have expressed himself in so undecided a manner. The opinion of Franklin adds perhaps little weight to a mathematical question, but it may tend to assist in lessening the repugnance which every true philosopher must feel to the necessity of embracing a physical theory different from that of Newton.

I have indeed been accused of insinuating 'that Sir Isaac Newton was but a sorry philosopher.' But it is impossible that an impartial person should read my essays on the subject of light without being sensible that I have as high a respect for his unparalleled talents and acquirements as the blindest of his followers, and the most parasitical of his defenders. I have acknowledged that 'his merits are great beyond all contest or comparison;' that 'his discovery of the composition of white light would alone have immortalized his name;' that the very arguments which tend to overthrow his hypothesis respecting the emanation of light, 'give the strongest proofs of the admirable accuracy of his experiments;' and that a person may, 'with the greatest justice, be attached to every doctrine which is stamped with the Newtonian approbation.' The printer of the 'Review,' feeling perhaps that the last expressions would militate too much in my favour, has thought fit to plunder me of them, by omitting the marks of quotation, and to attribute them to my antagonist. But, much as I venerate the name of Newton, I am not therefore obliged to believe that he was infallible. I see not with exultation, but with regret, that he was liable to err, and that his authority has, perhaps, sometimes even retarded the progress of science. It is now no longer denied that he was mistaken in an optical experiment respecting the dispersion of light; and the only attempt that has been made to explain the mistake merely shows that there was a possibility of his being misled by a singular combination of circumstances: in a case of mathematical optics he was certainly mistaken, as Dr. Smith has shown, when he asserted that a sphere of water produces a maximum of density in the light refracted at an angle of about 26°: in the mechanical estimation of force he erred when he calculated the precession of the equinoxes, and estimated the rotatory power of each particle of the earth's substance as simply proportional to its distance from the axis. These mistakes, and perhaps some others, have been acknowledged and corrected by later writers; other persons, less considerate, have attacked him where he was invulnerable. One of these is the gentleman whom I have reason to think the author of the remarks to which I am replying, and who, having first accused Newton of a palpable and fundamental blunder, appears now to be desirous of securing to himself the exclusive privilege of questioning his authority.

What I have hitherto said relates to the state of the question respecting the nature of light, as it stood before the publication of the first of the papers which have excited so much virulence. But I assert that this paper contains an argument sufficient to convert that which before was doubt and conjecture into probability and conviction. It was in May 1801 that I discovered, by reflecting on the beautiful experiments of Newton, a law which appears to me to account for a greater variety of interesting phenomena than any other optical principle that has yet been made known. I shall endeavour to explain this law by a comparison.

Suppose a number of equal waves of water to move upon the surface of a stagnant lake, with a certain constant velocity, and to enter a narrow channel leading out of the lake. Suppose then another similar cause to have excited another equal series of waves, which arrive at the same channel, with the same velocity, and at the same time with the first. Neither series of waves will destroy the other, but their effects will be combined: if they enter the channel in such a manner that the elevations of one series coincide with those of the other, they must together produce a series of greater joint elevations; but if the elevations of one series are so situated as to correspond to the depressions of the other, they must exactly fill up those depressions, and the surface of the water must remain smooth; at least I can discover no alternative, either from theory or from experiment.

Now I maintain that similar effects take place whenever two portions of light are thus mixed; and this I call the general law of the interference of light. I have shown that this law agrees, most accurately, with the measures recorded in Newton's 'Optics,' relative to the colours of transparent substances, observed under circumstances which had never before been subjected to calculation, and with a great diversity of other experiments never before explained. This, I assert, is a most powerful argument in favour of the theory which I had before revived: there was nothing that could have led to it in any author with whom I am acquainted, except some imperfect hints in those inexhaustible but neglected mines of nascent

R

inventions, the works of the great Dr. Robert Hooke, which had never occurred to me at the time that I discovered the law, and except the Newtonian explanation of the combinations of tides in the Port of Batsha.

It is unnecessary, on this occasion, to enter minutely into the consequences of the law of the interference of light : they have been the principal subjects of the three papers which have drawn down upon me the repeated anathemas of the self-erected Inquisition of the North. Not a single argument has been produced to invalidate it. The reviewer has cursorily observed that if the law were true, every surface opposed to the light of two candles would appear to be covered with fringes of colours. Let us suppose the assertion true—what will be the consequence? In all common cases the fringes will demonstrably be invisible; since, if we calculate the length and breadth of each fringe, we shall find that a hundred such fringes would not cover the point of a needle ; and an optician does not require to be told that a mixture like this constitutes a white light, not distinguishable by the senses from that which is supposed to have formed them.

In order to answer the charge of inconsistency in my opinions respecting the nature of light, I must begin by observing that there are two general methods of communicating knowledge : the one analytical, where we proceed from the examination of effects to the investigation of causes ; the other synthetical, where we first lay down the causes, and deduce from them the particular effects. In the synthetical manner of explaining a new theory we necessarily begin by assuming principles, which ought, in such a case, to bear the modest name of hypotheses; and when we have compared their consequences with all the phenomena, and have shown that the agreement is perfect, we may justly change the temporary term *hypothesis* into *theory*. This mode of reasoning is sufficient to attach a value and importance to our theory, but it is not fully decisive with respect to its exclusive truth, since it has not been proved that no other hypothesis will agree with the facts.

It is exactly in this manner that I have endeavoured to

proceed in my researches. By analysing the experiments of Newton, and comparing them with my own, I had arrived at principles, to which I gave, in my paper on the theory of light, the unassuming title of hypotheses; after comparing these principles with all the phenomena of light, and showing their perfect consistency, I thought myself authorised to make a conclusion, in my ninth proposition, which converts the hypothesis into a theory. I was justified in doing this, because no man had ever attempted to advance a theory which would bear to be compared mathematically with the phenomena that I enumerated. But, according to the nature of the only mode of reasoning which the circumstances allowed me, it was impossible to infer, from this synthetical comparison, that no other suppositions would agree with the phenomena ; and *I expressly remarked*, with respect to one of the four hypotheses which I laid down, that it was possible to find *others which might be substituted for it*. It is in this hypothesis and its consequences only, that I have since attempted to make any improvements. And such improvements I shall ever admit with pleasure, whether they arise from my own experiments, or from those of others. One immaterial correction of this kind I was obliged to make in consequence of Dr. Wollaston's most interesting observations upon the true division of the prismatic spectrum, which afford an additional proof that even Newton's experiments, frequently as they have been repeated by others, may sometimes stand in need of a more careful examination. And this modification, which has, in fact, little or no connexion with the essential parts of my theory, has been adduced as a proof of the 'fickle and vibratory nature of the medium that fills my mind.' The reviewer has indeed in another place denied the accuracy of Dr. Wollaston's experiment, but his objections are too futile to deserve notice.[1]

[1] In the following notice in the *Edinburgh Review* for April 1803, of his paper in the *Philosophical Transactions* for 1802, 'On the oblique Reflection of Iceland Crystal:'—'We were much disappointed to find that so acute and ingenious an experimentalist had adopted the wild optical theory of vibrations. After stating it, however, chiefly from Huygens, and applying it to explain the properties of the spar, he

Respecting another trifling change of sentiment, to which the reviewer has thought proper to attach great importance, I have hitherto abstained from explanation, in delicacy towards the gentleman whose observations were concerned; I wish to avoid insisting on his inaccuracy in a very easy calculation, and for the same reason I shall say nothing further on the subject at present.

When the reviewer asserts that 'a hypothesis is a work of fancy, useless in science,' it must be supposed that he is speaking of such hypotheses as have neither been originally deduced from experiments, nor afterwards compared with them: but when, in another of his articles, he condemns, as having impeded the progress of discovery, the beautiful hypothesis which has been applied, with the greatest success, by Aepinus, by Mr. Cavendish, and by Professor Robison, to the phenomena of electricity and magnetism, we can only regret that a person so void of a sense of physical elegance should have an opportunity of obtruding opinions like these on the public; and we may expect that he would say, if he dared, that even the hypothesis of

goes on to examine, by accurate experiments, whether the undulatory system agrees with the facts. The hypothesis is, that the different undulations of the elastic medium are spherical in almost all cases, but that, in the Iceland crystal, those undulations are spheroidal; and it must be acknowledged, the near coincidence of the experiments, which are extremely well contrived, and appear to be accurately conducted, give this theory a plausibility which it did not before possess. We would, however, remark that the hypothesis of Aepinus himself, by far the most consistent, simple, and universally applicable of any that has ever been proposed, is still only a gratuitous hypothesis; has acquired to its author only the praise of fanciful ingenuity; and has, perhaps, done more harm than good to the science of magnetism, by withdrawing the attention of philosophers from the patient and difficult, but profitable observation of nature, to the more easy but empty amusement of indulging their fancy.

'The hypothesis of Huygens is not, as Dr. Wollaston seems to think, the same with that of Euler and other unphilosophical inquirers. It approaches more nearly to that of Newton, and assumes the existence of an elastic medium, acting upon and influenced by the rays of light. These authors, misled by the nature of sound, do not admit the materiality of light, but assert that it is a vibration propagated through the medium. But ,short as these remarks are, we are loth to waste any more time on such a feeble and ill-conducted defence of an untenable and useless hypothesis.' Vol. ii. p. 99.—*Note by Dean Peacock.*

universal gravitation has presented an insuperable barrier to the advancement of experimental knowledge. He is at least determined to show that every hypothesis must be the work either of infancy or of dotage; and insinuates that the speculations which I have extracted from Newton's writings were merely the amusements of some vacant hours at the close of his scientific career. It is very true that the *queries* of Newton were given 'to the world' at a time when his brilliant and solid discoveries were fully established; but the papers which explain all his hypotheses concerning light the most at large, and to which I have had the most frequent occasion to refer, were read *to the Royal Society* more than ten years before he began to write his 'Principia;' and the principal reason that delayed their publication, appears to have been the apprehension of disputes with Dr. Hooke. Some were published in the 'Optics,' soon after Dr. Hooke's death; others are only to be found in Birch's History of the Royal Society. Had I not taken care to annex the dates to my quotations, the reviewer might easily have pleaded his ignorance in excuse for his misrepresentations.

The same plea of ignorance would be but an inadequate apology for the assertion of a positive falsehood, where he accuses me of referring to an *unpublished* work of my own. The reference could only be intended for the readers of the essay as a printed paper; my 'Syllabus' was published in January 1802; the 'Transactions' not till late in the spring; and if he had either sent to the publisher for this syllabus, or made inquiry for it among his literary friends even in Edinburgh, he might have found in it some information, on subjects which he appears to understand but imperfectly.

In the first paragraph of the review of my paper on the production of colours, the writer confesses that he has not '*sufficient fancy to discover*' how the 'interference of two portions of light' could ever produce an appearance of colour. The poverty of his fancy may indeed easily be admitted, but it is unfortunate that he either has not patience enough to read, or intellect enough to understand, the very papers that he is criticising; for, if he had perused with common attention my

Bakerian lecture on light, he might have understood such a production of colour without any exertion of fancy at all. He then quotes from me the assertion, that a 'black hair' does not produce the appearance of fringes, and he has even the modesty to refer to a certain page of my paper. I have there said, that a '*horse* hair' did not produce that appearance; and I have left it for the reviewer to decide whether the *horse* should be white or black. The truth is, that a fine wire, or a small hair, whether black or white, exhibits equally well the colours which I have described. If the fact were otherwise, it would be utterly unintelligible; for there is absolutely no foundation for the reviewer's insinuation, that any theory of these colours was deduced by De Dominis, or can be deduced by any other person, from the laws of refraction. He asserts that it is mathematically impossible for the light to bend round a hair. Grimaldi has long ago experimentally demonstrated this flexion, and called it diffraction: an effect which furnishes the most striking analogy between the motions of light and those of the waves of water.

The reviewer next complains of his utter *want of comprehension* of the difference between the colours of mixed plates, and those of the plates which have been described by Newton. Had he sufficiently studied the 'Optics' of Newton, he would have seen that the thickness of a simple plate of water must be only *three fourths* as great as that of a plate of air, in order to produce similar effects: in the colours which I have described, the thickness of the mixed plate was *six times* as great as that of the plate of air: the one series of rings *expanded*, upon inclining the plates, the other *contracted*. These distinctions are plain enough for any person of *ordinary* comprehension, and I was not aware that it was necessary to provide for *extraordinary* cases.

We are induced to suppose, from the page which immediately follows, that, to speak without a metaphor, neither the fancy nor the comprehension of the reviewer could enable him to distinguish a black spot from a white one. I have said, that when two glasses are brought into the most intimate contact possible, with the interposition of a certain fluid, the central spot of the rings of colours is nearly white : it was before known that, without any such interposition, the central spot, in similar circumstances, would be nearly black : and the critic sagaciously pronounces, that these effects are precisely the same. He quotes from Newton the expression of the 'pellucid central spot,' meaning the spot which reflected no light, and then explains it, as if it were exactly similar to that which, in my experiment, reflected nearly all the light that fell on it, and was therefore white.

That the lines which are quoted in the same page, from my paper, present, when thus insulated, an appearance of confusion and of vague reasoning is perhaps undeniable, and is perfectly excusable. The reviewer has not understood the paper in its entire state, and he might be sufficiently secure that his readers would never be able to extricate an intelligible sense from an arbitrary quotation of a few lines, taken out of the middle of a paragraph of connected reasoning. He misapprehends and misrepresents completely the whole subject of the explanation; he says that its object is to explain the blue colour of the lower part of the flame of a candle. Nothing was further from my thoughts than to assign any reason for this blueness : what I attempted to illustrate, was an original and important observation made by Dr. Wollaston, that a portion of the blue flame of a candle appeared, when viewed through a prism, to be divided into a number of distinct masses or images. My illustration of this phenomenon has not the slightest connection with what the reviewer calls his solution of the appearance of different colours in different flames, which he so humbly intreats his readers to compare with it. I am not therefore obliged to give an opinion of any kind respecting this pretended explanation of a phenomenon foreign to the subject; if I were, it would be sufficient to say, that no such laws could be supposed to operate, upon the principles of mechanical forces, without producing different velocities in light of different colours. But the passage fortunately affords me a most convincing proof of the nature of the source from which this torrent of invective has originated. We are here told, that the doctrine of the different *flexibility* of

light is now *universally admitted.* I have searched into all the
works that I could find in the libraries to which I have had
access, for opinions respecting the nature of light, and, as far as
I have discovered, the different flexibility of light is *admitted,*
in the absurd and unwarrantable sense in which it is here
employed, *by three writers only.* The first is Mr. Henry
Brougham, the second the anonymous author of an article
in the 'Encyclopædia Britannica,' and the third the assailant
whose injurious attacks I am now repelling. From so remarkable a coincidence, I think myself authorised to conclude, that
these three writers are one and the same. I have before
hinted that Mr. Brougham's doctrines have been sufficiently
confuted, by Professor Prevost of Geneva.[1] Mr. Prevost has
satisfactorily defended the experiments of Newton from the
imputations of Mr. Brougham; but in other respects he has
perhaps treated the young theorist with too much lenity.

I have now answered everything that was intended as an
argument, in the articles published in the second number of the
'Review.' This constitutes, in fact, but a small part of those
articles: they have much less the appearance of the impartial
discussion of a long disputed question in natural philosophy,
than of the buffoonery of a theatrical entertainment, or of the
jests of a pert advocate, endeavouring to place in a ridiculous
light the evidence of his adversary. To answer such an attack
in similar language would be degrading; to attempt to oppose
it by argument would be futile. I shall refrain, therefore, from
noticing any of the additional scurrilities which have been
copiously intermixed by the same writer with his remarks
on my last paper. I say the same, because I am unwilling to
suppose that this island has produced two persons capable of so
stupidly misundertanding, and so wilfully misrepresenting.
But their identity is of no consequence to the discussion, and it
is unnecessary to inquire for proofs of it. The whole purpose

---

[1] In the *Philosophical Transactions* for 1798, vol. lxxxviii. p. 321
in a paper entitled 'Some Optical Remarks chiefly relative to the Re-
flexibility of the Rays of Light.'—*Note by Dean Peacock.*

of the paper inserted in the ninth number of the 'Review' might be supposed to have been, not to confute the principles which the writer attacks, but to show that he is incapable of understanding even the simplest of them.

I have asserted that two series of *undulations, interfering* with each other at certain relative *intervals*, necessarily produce certain modifications in their joint effects. These terms not only belong to the same theory, but are parts of the same position which I have already illustrated by a familiar comparison in these remarks. The author of the critique has sagaciously observed, that 'they who object to the theory of interference, have only to *turn a page*, and they find the theory of intervals, and they need but go on *a section further*, and the vibrations and undulations are very much at their service.'

This specimen is sufficient to explain how naturally it must appear to him 'unaccountable,' that the process of interference should produce certain effects, some of which I never supposed that it could produce, and others which none who rightly understood my theory could ever doubt that it must produce. He asks, 'on what known principle' can the production of coloured fringes from two beams of white light be explained? I answer, certainly on no principle that was known before; but upon consideration of the law which I have discovered, most simply and unavoidably.

The reviewer has afforded me, in the next observation, an opportunity for a triumph as gratifying as any triumph can be where the enemy is so contemptible. Conscious of inability to explain the experiment which I have advanced, too ungenerous to confess that inability, and too idle to repeat the experiment, he is compelled to advance the supposition that it was incorrect, and to insinuate that my hand may easily have erred through a space so narrow as one-thirtieth of an inch. But the truth is, that my hand was not concerned: the screen was placed on a table, and moved mechanically forwards with the utmost caution. The experiment succeeded in some circumstances where the breadth of the object was doubled or tripled; and I assert that it was as easy to me to estimate an interval of one-thirtieth of an inch, as an interval a hundred or a thousand times as great. Let him make the experiment, and then deny the result if he can.

With equal pertinacity of blundering, he has remarked that the interference of light inflected by two continuous edges ought, upon my principles, to produce not continued fringes, but only 'square or rectangular spots of fringe.' Was it not enough to have demonstrated the weakness of his powers with regard to physical laws? And was it necessary to induce his readers to suppose him incapable of going through a little algebraical calculation leading to the properties of the hyperbola? Let right lines be inflected from the edges of a rectangular object into its shadow, so as to cut off portions with the opposite lines, exceeding their own length by a given interval, and I maintain that the intersections will form continued curves, and that those curves will be hyperbolas: the shape of the fringes ought *not*, therefore, to be that of detached spots, but of hyperbolical curves.

It is 'a metaphysical absurdity,' says the reviewer, to assert that qualities can 'move' in concentric surfaces. I have not said that the qualities of light 'move' in concentric surfaces, but that they 'succeed each other' in concentric surfaces; and in this there is certainly no metaphysical absurdity. Condensation and rarefaction are qualities of the air, and it will not be denied that, in every musical sound, condensation and rarefaction continually succeed each other in concentric surfaces.

Upon my train of argument respecting the nature of light, the reviewer observes, first, that an analogy is made the ground of an inference. I answer, that when the analogy is sufficiently close, it is a most satisfactory ground of physical inference. Secondly, he says that a gratuitous assumption is set down as a necessary truth. I reply, that the assumption is not gratuitous; that nobody, except for the sake of argument, will deny, or can deny it; should it be denied, it would be perfectly easy to substantiate it by showing the unavoidable contradictions that would result from any alternative that could be substituted for it. The remaining part of the paragraph is as correctly quoted as that edition of the Bible was printed, in which the only error was the omission of the word *not* in the seventh Commandment: here the monosyllabie *but*, which completely inverts the sense of the passage, and which would have entirely destroyed the force of the criticism, is therefore very prudently omitted.

I have inserted a caution relating to deceptions in the examination of microscopical objects, not in order to attach any additional merit to my own explanations, but as a hint naturally arising out of the subject. The same caution might perhaps have been suggested by the results of some former experiments, but the particular appearances that would be produced by such fallacies could never before have been so minutely indicated. That the images of very small objects on the retina may possibly be affected by such causes, is the natural inference from my principles; and it is of no consequence to this position whether the reviewer can or cannot explain them from his own.

My comparison of a grove of trees pervaded by the wind, with the particles of a material body, separated, as all modern philosophers have supposed them to be, by intervals incomparably greater than their diameters, and allowing an inconceivably rare medium to penetrate with perfect freedom every interstice, could scarcely have appeared obscure or inapplicable to any man unblinded by prejudice or unbiassed by malevolence.

I have already said enough of Newton to show how I venerate his character, as the first of mathematicians and the greatest of philosophers. Perhaps, however, the mention of persons whose views are ' still less enlarged ' than his own, may imply in some measure what I never intended, and may therefore require some little apology, especially as the expressions will bear to be applied to the objections which I am now endeavouring to refute. It was, indeed, a want of respect to his illustrious memory to place the superficial and dogmatical fancies of a writer in the 'Edinburgh Review' in any kind of comparison with the deep and refined imaginations of a Newton. Instead

of ' still less enlarged' and enlightened, I ought to have called
them narrow and confused, selfish and interested, puerile and
ostentatious.

The indignation of the same violent and arbitrary tribunal
has been excited and called forth by a declaration from a man
whose approbation is so much the more valuable as it is always
bestowed with the most cautious regard to experimental accuracy and logical induction.   Dr. Wollaston has observed that
' the theory of Huygens affords, as has lately been shown by
Dr. Young, *a simple explanation of several phenomena not yet
accounted for by any other hypothesis.*'   His own observations
on Iceland crystal accord throughout, he says, with this hypothesis of Huygens; the measures that he has taken ' correspond
more nearly than could well happen to a false theory.'   But
he contents himself with stating these undeniable facts; and
the reviewer goes too far when he asserts that Dr. Wollaston
' has adopted the wild optical theory of vibrations.'   If Dr.
Wollaston had then been acquainted with the experiments and
calculations which I have made since that time, it is possible
that his assent might have been much more complete and un-
reserved.   But while I allow to his experiments all the merit
that a clear conception, a vigorous mind, a steady hand, and
an accurate eye can bestow on them, it must not be said by
the Edinburgh Reviewers that his experiments have given the
theory ' a plausibility which it did not before possess.'   As experiments, they have all the merit of originality, for the author,
when he made them, was unacquainted with those of Huygens;
and his most ingenious invention of an instrument for measuring refractive powers enabled him, with great ease, to improve
and extend them.   But the experiments of Huygens were
elaborate and diversified, and every argument that can be inferred from Dr. Wollaston's observations had been anticipated
by this great philosopher upon the ground of his own.   It is
true that our reviewer was not likely to have troubled himself
with Huygens's treatise of light; his business is to censure
others, and not to inform himself; it was easier for him to call
this doctrine 'a clumsy hypothesis,' and 'a dull invention,' than

to investigate its truth and to admire its elegance. He has indeed made distinctions between Huygens's doctrine and mine, which serve but to prove still more strongly that he was acquainted with neither; I shall only answer his epithets by a quotation from a writer, whose merits the testimony of Newton is well known to have raised far above the ordinary rank of his contemporaries. In Cotes's lectures on hydrostatics, where he is speaking of the velocity of light, and takes occasion to mention the hypothesis of Huygens, the following passage occurs :—'When we take a particular view of the several parts of this hypothesis, it appears to be so very ingeniously contrived, and so handsomely put together, that one can hardly forbear to wish it were true.' The evidence was at that time imperfect, but the symmetry was complete.

The reviewer has thought proper to unite, in several instances, with his invectives against me, some ridicule of the objects of the Royal Institution of Great Britain ; an institution in which its managers have studied to concentrate all that is useful in science, or elegant in literature. This connexion appears to him to add so much weight to his arguments, that he has chosen, without further provocation, to insinuate its existence more than a year after it had been dissolved. I accepted the appointment of Professor of Natural Philosophy in the Royal Institution as an occupation which would fill up agreeably and advantageously such leisure hours as a young practitioner of physic must expect to be left free from professional cares. I was led to hope that I should be able to impress an audience formed of the most respectable inhabitants of the metropolis, with such a partiality as the moderately well-informed are inclined to entertain, for those who appear to know even a little more than themselves of matters of science ; that I might be of use to the public in disseminating the true principles of natural philosophy; and that I might in future be remunerated by the enjoyment of a more extensive confidence in my professional abilities than could have been granted to a person less generally known. While I held the situation, I wished to make my lectures as intelligible as the nature of the subjects permitted; but I must confess that it was not my ambition to render them a substitute for those of any superficial experimenter, that was in the habit of delivering courses of natural philosophy for the amusement of boarding schools. Whatever may have been the imperfections of my lectures, it cannot be asserted, except perhaps in the ' Edinburgh Review,' that they were fit for audiences of ladies of fashion only. After fulfilling, for two years, the duties of the Professorship, I found them so incompatible with the pursuits of a practical physician, that, in compliance with the advice of my friends, I gave notice of my wish to resign the office. I think it, however, just to the Institution, to the public, and to myself, that the result of my labours, throughout the whole extent of natural philosophy and the mechanical arts, should be rendered of some permanent utility; and I have since collected such a mass of further references to works of all ages and of all nations, accompanied by many notes and extracts from them, that it will henceforwards be easy for every student and every author to know at once what has been done, and what remains to be done, in the subject of his particular researches; and to what books he must apply for the best information ; where further information is required, and can be obtained. Considering how widely this information is at present scattered, I trust that I shall have rendered a service of some importance to every department of the sciences, and I am now on the point of preparing my book for immediate publication. With this work my pursuit of general science will terminate : henceforwards I have resolved to confine my studies and my pen to medical subjects only. For the talents which God has not given me I am not responsible, but those which I possess I have hitherto cultivated and employed as diligently as my opportunities have allowed me to do; and I shall continue to apply them with assiduity, and in tranquillity, to that profession which has constantly been the ultimate object of all my labours.[1]

[1] 'Of the preceding most masterly Reply,' says Dean Peacock, in his excellent Life of Young, 'which was published in the form of a pamphlet,

Young has been defended with great ability and out-spokenness by Dr. Whewell and Dean Peacock. I content myself here with placing the assailant and the assailed side by side. It is not necessary to add a word to strengthen the verdict which, throughout all time, will be pronounced upon the writer whose invective drew from Dr. Young the foregoing vindication.

MEASUREMENT OF THE WAVES OF LIGHT.

THE diffraction fringes described in Lecture II., instead of being formed on the retina, may be formed on a screen, or upon ground glass, when they can be looked at through a magnifying lens from behind, or they can be observed in the air when the ground glass is removed. Instead of permitting them to form on the retina, we will suppose them formed on a screen. This places us in a condition to understand, even without trigonometry, the solution of the important problem of measuring *the length* of a wave of light.

We will suppose the screen so distant that the rays falling upon it from the two margins of the slit are sensibly parallel. We have learned in Lecture II. that the first of the dark bands corresponds to a difference of marginal path of one undulation; the second dark band to a difference of path of two undulations; the third dark band to a difference of three undulations, and so on. Now the angular distance of the bands from the centre is capable of exact measurement; this distance depending, as already stated, on the width of the slit. With a slit 1·35[1] millimeters wide, Schwerd found the angular distance of the first dark band from the centre of the field to be 1′ 38″; the angular distances of the second, third, fourth dark band being twice, three times, four times this quantity.

it was stated by its author, that *one copy only was sold*: it consequently produced no effect in vindicating his scientific character, or in turning the current of public opinion in favour of his theory.'

[1] The millimeter is about $\frac{1}{25}$th of an inch.

Let A B, fig. 59, be the plate in which the slit is cut, and C D the grossly exaggerated width of the slit, with the beam of red light proceeding from it at the obliquity corresponding to the first dark band. Let fall a perpendicular from one edge, D, of the slit on the marginal ray of the other edge at *d*. The distance, C *d*, between the foot of this perpendicular

FIG. 59.

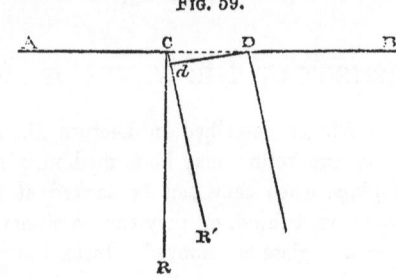

and the other edge is the length of a wave of the light. The angle C D *d*, moreover, being equal to R C R', is, in the case now under consideration, 1' 38". From the centre D, with the width D C as radius, describe a semicirle; its radius D C being 1·35 millimeters, the length of this semicircle is found by an easy calculation to be 4·248 millimeters. The length C *d* is so small that it sensibly coincides with the arc of the circle. Hence the length of the semicircle is to the length C *d* of the wave as 180° to 1' 38", or, reducing all to seconds, as 648,000" to 98". Thus, we have the proportion—

648,000 : 98 :: 4·248 to the wave-length C *d*.

Making the calculation, we find the wave-length for this particular kind of light to be 0·000643 of a millimeter, or 0·000026 of an inch.

WATER CRYSTALLIZATION.

THE following letter from my excellent friend Professor Joseph Henry refers to the surprising case of crystallization figured in the frontispiece, and for which I am indebted to the kindness of Professor Lockett :—

'Smithsonian Institution, Washington:
'March 24, 1873.

'MY DEAR PROFESSOR TYNDALL,—Accompanying this I send you a photograph, at the request of Professor S. H. Lockett, of the Louisiana State University, of which the following is his explanation :—

' "In my drawing room I kept a wash-basin in which to rinse out the colour from my water-colour brushes. This colour gradually formed a uniform sediment of an indefinite tint over the bottom of the basin. On the night of the 26th of December last, which was an unusually cold one for this climate, the water in the basin froze. On the melting of the ice the next day, the beautiful figure you see on the photographs was left in the sediment. I carefully poured the water from the basin, let the sediment dry, and thus perfectly preserved the figure. It has been accurately photographed by an artist in this city. The negative is preserved, and if you would like to have any more copies they can readily be obtained.

' "We are not much accustomed in this warm country of ours to the beautiful ' forms of water,' and this has struck me as a little remarkable, and worthy of being kept."

' The fact that the results have been produced by coloured sediment indicates a method of exhibiting the effects of crystallization in an interesting manner.

' JOSEPH HENRY,

' Secretary, Smithsonian Institution.'

S

LIFE AND CRYSTALLIZATION.

IN Lecture III. the phenomena of crystallization are referred
to, and their relationship to the phenomena of life hinted at.
In the opening address to the mathematical and physical
section of the British Association at Norwich, I sought to
give clear expression to the notions I then entertained upon
this subject.   The following extract, however, from a journal,
written at Dinan in 1855, illustrates the common fact that
thoughts expressed in riper years have been the substantial
possession of our earlier ones :—

'At one or two points, where the view was more than
usually beautiful, the rug was spread on the grass, and we sat
down.   H. started the question whether natural beauty tended
to make men better, and he referred to the tropics in proof
that such beauty did not always lead to beauty of life.   I
contended that to examine the effect of natural beauty it must
be detached from other influences and its effect studied
alone.   If people of the same character and capacities, and
possessing the same facilities of obtaining the necessaries of
life, were placed, some amid beautiful scenery, and others
amid scenery dreary and dull, the former would have the
advantage.   H. contended that culture was necessary before
the mind could extract benefits from fine scenery, while I
urged that the culture which he demanded would itself be
promoted by a pleasant scene.

' The influence of climate upon man led us to consider solar
influence generally in relation to organic growth.   A large
elm was at hand.   It was certainly a mechanical act to lift
the matter of that tree so high in opposition to the force of
gravity.   Against this power the molecules had ascended ;
unaided by this power they had diverged into branches and
spread themselves into innumerable leaves.   Now the
molecules here had been either lifted by a power external to
themselves or by a power resident in themselves.   No physical

philosopher can accept the former notion; the latter must therefore be assumed. Consider from this point of view the experience of the present year (1855). For weeks longer than usual a low temperature had prevented the vegetable world from giving any signs of life; but at length the sun gained power, life as a consequence awoke, and it was still working to augment the beauty around us.

'But what is the thing which we here call life? and how is it that light and heat can thus affect it? The answer to this presupposes an answer to the other question, What are light and heat? Near the elm tree stood a birch, with its tremulous leaves flapping in the morning air. Here was motion, but it was not the motion to which we gave the name of life. Each leaf in this case moved as a mass, whereas life required an internal motion of the molecules. How are we to figure this? Suppose the leaves gifted with attractive and repulsive forces, suppose them moreover withdrawn from the action of gravity, and abandoned to their own interaction. To fix the ideas, suppose the point of each leaf to repel the points of all other leaves and to attract the other extremities, and the root of each leaf to repel all other roots but the points. A number of such leaves brought together and permitted to act upon each other would arrange themselves in a particular manner, assuming finally a position in which the forces would be in equilibrium. When thus at rest let us suppose the breeze which now causes them to quiver to act upon them, to ruffle them—in a word, to disturb the pre-existing equilibrium. There would be a constant effort on the part of the leaves to restore it, and in making this effort they would pass through different arrangements, the entire mass of leaves assuming different shapes in the passage from one arrangement to another. By this rough image we may help ourselves to a conception of those processes to which we give the name of life.

'The ultimate particles, or molecules, of matter are endowed with forces typified by those here ascribed to the leaves. Under the operation of such forces the molecules of the seed assume positions of rest, in which they would remain for ever if undisturbed by an external force. But a source of disturbance exists in the heat of the sun, which comes to us in undulations through the ether of space. On the mutually locked molecules of the seed the vibrations eventually impinge; they and the matter surrounding them are thrown into motion, interaction follows, an effort to restore the disturbed equilibrium is immediately exerted, but incessantly defeated, and the molecular struggle results in the formation of the tree. *Life is therefore scientifically defined as an incessant effort to restore a disturbed equilibrium.*

'These speculations spring out of no profane curiosity; they are inevitably forced upon the profoundly thinking mind. Let us indulge in them with reverence, but also with courage; for though in thus acting we may cause many of the idols of our youth to vanish, and reduce many mysteries to mechanics, we shall assuredly in the end leave the miracle of nature unimpaired. There is no escape, I say, from thoughts like these. Unless we assign to each particular plant an architect who lifts the molecules and places them in position, the physical processes of life must be due to the operation of the forces wherewith the molecules are endowed.

' Who is the builder in the case of a crystal [of the plumes in our frontispiece for example] ? Either a detached architect does the business, or these wonderful structures are self-erected, in virtue of their inherent forces. In building a crystal nature makes her first real effort as an architect. Here we have the first gropings of the so called vital force; but the most wonderful manifestations of this force, though depending on processes of higher complexity, are, I hold, of the same quality as those concerned in the growth of a crystal.

' Will the poet or the imaginative man shrink from these notions as cold and mechanical ? Why should he ? For what have we done but pushed the eternal mystery a little farther back ? We reduce life to the operation of molecular forces; but how came the molecules to be thus endowed ? Who or what gave to these forces their particular tendencies and direction ?

Let us contemplate that cycle of operations in which the seed produces the plant, the plant the flower, and the flower the seed again, thus returning with the unerring fidelity of a planet in its orbit to the point from which it started. All these processes are undoubtedly due to the action of molecular forces. But who or what planned their manner of action? Who or what endowed them with the power of taking up at a given time a determinate position, to be followed by another and another through the course of ages? Yonder butterfly has a spot of orange on its wing; if we look into a book written a hundred years ago, where that butterfly is figured, we find the selfsame spot upon the wing. Now the spot depends purely on the manner in which the light falling on and entering the wing is discharged from it, and this again depends upon the molecular structure of the wing. For a century, then, the molecules have gone through successive cycles; butterflies have been begotten, have grown, and died, and still we find the architecture the same. Is not this amazing? And what is the explanation? We may have a thousand proximate reasons, but at bottom we have no explanation. Still we stand firm within our range. There is nothing in the architecture of that wing which may not yet find its Newton to show that the laws and principles brought into play in its construction are qualitatively the same as those brought into play in the construction of the solar system. There is no essential distinction between organic and inorganic; the forces present in the one, when duly compounded, can and must produce the phenomena of the other.

'Thus far do I proceed with absolute confidence; and I am ready to take a step farther. The brain of man itself is assuredly an assemblage of molecules, arranged according to physical laws; but if you ask me to deduce from this assemblage the least of the phenomena of sensation or thought, I lay my forehead in the dust and acknowledge human helplessness. Here speculation folds her wings, for beyond this point there is no medium to sustain her flight.'

ON THE SPECTRA OF POLARIZED LIGHT.

Mr. William Spottiswoode has recently introduced to the members of the Royal Institution, in a very striking form, a series of experiments on the spectra of polarized light. With his large Nicol's prisms he first repeated and explained the experiments of Foucault and Fizeau, and subsequently enriched the subject by very beautiful additions of his own. I here append a portion of the abstract of his discourse :—

'It is well known that if a plate of selenite sufficiently thin be placed between two such Nicol's prisms, or, more technically speaking, between a polarizer and analyzer, colour will be produced. And the question proposed is, What is the nature of that colour? is it simply a pure colour of the spectrum, or is it a compound, and if so, what are its component parts? The answer given by the wave theory is in brief this: In its passage through the selenite plate the rays have been so separated in the direction of their vibrations and in the velocity of their transmission, that, when re-compounded by means of the analyzer, they have in some instances neutralized one another. If this be the case, the fact ought to be visible when the beam emerging from the analyzer is dispersed by the prism; for then we have the rays of all the different colours ranged side by side, and if any be wanting, their absence will be shown by the appearance of a dark band in their place in the spectrum. But not only so; the spectrum ought also to give an account of the other phenomena exhibited by the selenite when the analyzer is turned round, viz. that when the angle of turning amounts to 45° all trace of colour disappears; and also that when the angle amounts to 90° colour reappears, not, however, the original colour, but one complementary to it.

' You see in the spectrum of the reddish light produced by

the selenite a broad but dark band in the blue; when the analyzer is turned round the band becomes less and less dark, until when the angle of turning amounts to 45° it has entirely disappeared.   At this stage each part of the spectrum has its own proportional intensity, and the whole produces the colourless image seen without the spectroscope. Lastly, as the turning of the analyzer is continued, a dark band appears in the red, the part of the spectrum complementary to that occupied by the first band ; and the darkness is most complete when the turning amounts to 90°.   Thus we have from the spectroscope a complete account of what has taken place to produce the original colour and its changes.

'It is further well known that the colour produced by a selenite, or other crystal plate is dependent upon the thickness of the plate.   And, in fact, if a series of plates be taken, giving different colours, their spectra are found to show bands arranged in different positions.   The thinner plates show bands in the parts of the spectrum nearest to the violet, where the waves are shorter, and consequently give rise to redder colours; while the thicker show bands nearer to the red, where the waves are longer, and consequently supply bluer tints.

'When the thickness of the plate is continually increased, so that the colour produced has gone through the complete cycle of the spectrum, a further increase of thickness causes a reproduction of the colours in the same order; but it will be noticed that at each recurrence of the cycle the tints become paler, until when a number of cycles have been performed, and the thickness of the plate is considerable, all trace of colour is lost.   Let us now take a series of plates, the first two of which, as you see, give colours; with the others which are successively of greater thickness the tints are so feeble that they can scarcely be distinguished.   The spectrum of the first shows a single band ; that of the second, two ; showing that the second series of tints is not identical with the first, but that it is produced by the extinction of two colours from the components of white light.   The spectra of the others show series of bands more and more numerous in proportion to the thickness of the plate, an array which may be increased indefinitely. The total light, then, of which the spectrum is deprived by the thicker plates is taken from a greater number of its parts ; or, in other words, the light which still remains is distributed more and more evenly over the spectrum; and in the same proportion the sum total of it approaches more and more nearly to white light.

' These experiments were made more than thirty years ago by the French philosophers, MM. Foucault and Fizeau.

' If instead of selenite, Iceland spar, or other ordinary crystals, we use plates of quartz cut perpendicularly to the axis, and turn the analyzer round as before, the light, instead of exhibiting only one colour and its complementary with an intermediate stage in which colour is absent, changes continuously in tint; and the order of the colour depends partly upon the direction in which the analyzer is turned, and partly upon the character of the crystal, *i.e.* whether it is right-handed or left-handed. If we examine the spectrum in this case we find that the dark band never disappears, but marches from one end of the spectrum to another, or *vice versâ*, precisely in such a direction as to give rise to the tints seen by direct projection.

' The kind of polarization effected by the quartz plates is called circular, while that effected by the other class of crystals is called plane, on account of the form of the vibrations executed by the molecules of æther; and this leads us to examine a little more closely the nature of the polarization of different parts of these spectra of polarized light.

' Now, two things are clear : first, that if the light be plane-polarized, that is, if all the vibrations throughout the entire ray are rectilinear and in one plane, they must in all their bearings have reference to a particular direction in space, so that they will be differently affected by different positions of the analyzer. Secondly, that if the vibrations be circular, they will be affected in precisely the same way (whatever that may be) in all positions of the analyzer. This statement merely recapitulates a fundamental point in polarization. In fact, plane-polarized light is alternately transmitted and extinguished by the analyzer as it is turned through 90°; while circularly-polarized light [if we could get a single ray] remains to all appearance unchanged. And if we examine carefully the spectrum of light which has passed through a selenite, or other ordinary crystal, we shall find that, commencing with two consecutive bands in position, the parts occupied by the bands and those midway between them are plane polarized, for they become alternately dark and bright; while the intermediate parts, *i.e.* the parts at one-fourth of the distance from one band to the next, remain permanently bright. These are, in fact, circularly polarized. But it would be incorrect to conclude from this experiment alone that such is really the case, because the same appearance would be seen if those parts were unpolarized, *i.e.* in the condition of ordinary lights. And on such a supposition we should conclude with equal justice that the parts on either side of the parts last mentioned (*e.g.* the parts separated by eighth parts of the interval between two bands) were partially polarized. But there is an instrument of very simple construction, called a " quarter-undulation plate," a plate usually of mica, whose thickness is an odd multiple of a quarter of a wave length, which enables us to discriminate between light unpolarized and circularly polarized. The exact mechanical effect produced upon the ray could hardly be explained in detail within our present limits of time; but suffice it for the present to say that when placed in a proper position, the plate transforms plane into circular and circular into plane polarization. That being so, the parts which were originally banded ought to remain bright, and those which originally remained bright ought to become banded during the rotation of the analyzer. The general effect to the eye will consequently be a general shifting of the bands through one-fourth of the space which separates each pair.

' Circular polarization, like circular motion generally, may of course be of two kinds, which differ only in the direction of the motion. And, in fact, to convert the circular polarization produced by this plate from one of these kinds to the other (say from right-handed to left-handed, or *vice versâ*), we have only to turn the plate round through 90°. Conversely right-handed circular polarization will be changed by the plate into plane polarization in one direction, while left-handed will be changed into plane at right angles to the first. Hence, if the plate be turned round through 90° we shall see that the bands are shifted in a direction opposite to that in which they were moved at first. In this therefore we have evidence not only that the polarization immediately on either side of a band is circular; but also that that immediately on the one side is right-handed, while that immediately on the other is left-handed. [1]

'If time permitted, I might enter still further into detail, and show that the polarization between the plane and the circular is elliptical, and even the positions of the longer and shorter axes and the direction of motion in each case. But sufficient has, perhaps, been said for our present purpose.

'Before proceeding to the more varied forms of spectral bands, which I hope presently to bring under your notice, I should like to ask your attention for a few minutes to the peculiar phenomena exhibited when two plates of selenite giving complementary colours are used. The appearance of the spectrum varies with the relative position of the plates. If they are similarly placed—that is, as if they were one plate of crystal—they will behave as a single plate, whose thickness is the sum of the thicknesses of each, and will produce double the number of bands which one alone would give; and when the analyzer is turned, the bands will disappear and re-appear in their complementary positions, as usual in the case of plane-polarization. If one of them be turned round through 45°, a single band will be seen at a

[1 At these points the two rectangular vibrations into which the original polarized ray is resolved by the plates of gypsum act upon each other like the two rectangular impulses imparted to our pendulum in Lecture IV., one being given when the pendulum is at the limit of its swing. Vibration is thus converted into rotation.]

particular position in the spectrum. This breaks into two, which recede from one another towards the red and violet ends respectively, or advance towards one another according to the direction in which the analyzer is turned. If the plate be turned through 45° in the opposite direction, the effects will be reversed. The darkness of the bands is, however, not equally complete during their whole passage. Lastly, if one of the plates be turned through 90°, no bands will be seen, and the spectrum will be alternately bright and dark, as if no plates were used, except only that the polarization is itself turned through 90°.

'If a wedge-shaped crystal be used, the bands, instead of being straight, will cross the spectrum diagonally, the direction of the diagonal (dexter or sinister) being determined by the position of the thicker end of the wedge. If two similar wedges be used with their thickest ends together, they will act as a wedge whose angle and whose thickness is double of the first. If they be placed in the reverse position they will act as a flat plate, and the bands will again cross the spectrum in straight lines at right angles to its length.

'If a concave plate be used the bands will dispose themselves in a fanlike arrangement, their divergence depending upon the distance of the slit from the centre of concavity.

'If two quartz wedges, one of which has the optic axis parallel to the edge of the refractory angle, and the other perpendicular to it, but in one of the planes containing the angle (Babinet's Compensator), the appearances of the bands are very various.

' The diagonal bands, beside sometimes doubling themselves as with ordinary wedges, sometimes combine so as to form longtitudinal (instead of transverse) bands; and sometimes cross one another so as to form a diaper pattern with bright compartments in a dark framework, and *vice versâ*, according to the position of the plates.

' The effects of different dispositions of the interposed crystals might be varied indefinitely; but enough has perhaps been said to show the delicacy of the method of spectrum analysis as applied to the examination of polarized light.'

The singular and beautiful effect obtained with a circular plate of selenite, thin at the centre, and gradually thickening towards the circumference, is easily connected with a similar effect obtained with Newton's rings. Let a thin slice of light fall upon the glasses which show the rings, so as to cover a narrow central vertical zone passing through them all. The image of this zone upon the screen is crossed by portions of the iris rings. Subjecting the reflected beam to prismatic analysis, the resultant spectrum may be regarded as an indefinite number of images of the zone placed side by side. In the image before dispersion we have *iris-rings*, the extinction of the light being nowhere complete; but when the different colours are separated by dispersion, each colour is crossed transversely by its own system of dark interference bands, which become gradually closer with the increasing refrangibility of the light. The complete spectrum, therefore, appears furrowed by a system of continuous dark bands, crossing the colours transversely, and approaching each other as they pass from red to blue.

In the case of the plate of selenite, a slit is placed in front of the polarizer, and the film of selenite is held close to the slit, so that the light passes through the central zone of the film. As in the case of Newton's rings, the image of the zone is crossed by iris-coloured bands; but when subjected to prismatic dispersion, the light of the zone yields a spectrum furrowed by bands of complete darkness exactly as in the case of Newton's rings, and for a similar reason. This is the beautiful effect described by Mr. Spottiswoode as the fan-like arrangement of the bands—the fan opening out at the red end of the spectrum.

INDEX.

LONDON: PRINTED BY
SPOTTISWOODE AND CO., NEW-STREET SQUARE
AND PARLIAMENT STREET